CONSCIOUSNESS AND MENTAL LIFE

DANIEL N. ROBINSON

Consciousness AND Mental Life

Columbia University Press New York

Columbia University Press
Publishers Since 1893
New York Chichester, West Sussex

Library of Congress Cataloging-in-Publication Data
Robinson, Daniel N., 1937–
Consciousness and mental life / Daniel N. Robinson.
p. cm.
Includes bibliographical references and index.
ISBN 978-0-231-14100-0 (cloth : alk. paper) —
ISBN 978-0-231-51280-0 (e-book)
1. Consciousness. 2. Cognitive neuroscience. I. Title.

B105.C477R63 2008
126—dc22 2007019372

Columbia University Press books are printed on
permanent and durable acid-free paper.
This book is printed on paper with recycled content.

Printed in the United States of America
Designed by Audrey Smith

c 10 9 8 7 6 5 4 3 2 1

CONTENTS

Preface vii
Acknowledgments xi

1. The Greeks (Again) and the "Consciousness" Problem 1
2. The Problem of Consciousness "Solved" 17
3. "Cartesianism" Revisited 51
4. Higher-Order Thought: A Machine in the Ghost 83
5. Self-Consciousness 101
6. Emotion 145
7. Motives, Desires, and Fulfillment 169
8. Plans: An Epilogue 201

Notes 211
Index 233

PREFACE

In recent decades, each of them curiously referred to as "the decade of the brain," the more venerable issues in philosophy and psychology have been awkwardly absorbed into a scientific framework variously identified as "brain science," "cognitive science," or "cognitive neuroscience." The literature generated by this perspective is now vast and growing, ranging from highly technical monographs to folksy invitations to witness the "Revolution." And, of course, all agree that there has been some sort of revolution.

My aims with this slim volume are modest. First, I hope to convey to readers just what it is about the topics covered that has made them problematic within the councils of thought for many centuries. In doing so, I must also make clear that earlier thinkers, some of whom surely rank among the greatest minds of history, were not feckless incompetents now to be seen as the source of conceptual blunders readily detected by first-year philosophy students. Descartes, for example, was at least clever enough to save himself from something called "Cartesianism" and its alleged error.

However, over the course of that long debate on the nature of human nature (and its relation to the balance of nature), there have been nontrivial developments, both perspectival and technical. These have led inevitably to seasonal shifts in attention, in emphasis, and in the very criteria of explanation and discovery. Accordingly, it is instructive to examine how the reigning perspectives of an age come to identify topics worthy of study and establish the grounds on which progress may be claimed. In our own time, achievements in the physical sciences and engineering have bolstered confidence that the same methods and models are applicable across the board. Thus progress is perceived when (and *solely* when) a traditional issue is reformulated in terms compatible with (what is taken to be) physics. Of course, what is taken to be physics is itself not uncontroversial or static, so even adherents of this doctrine must remain as alert as they are typically defensive.

There is perhaps a more fundamental aim that the following pages seek to realize. The topic addressed here is one about which every reader might claim authoritative knowledge. But "authority" comes in different forms. It is entirely possible for there to be significant achievements *of a sort* in our understanding of consciousness, but none at all in other and different senses of both "understanding" and "consciousness." An example from biology is useful. The Krebs cycle explains the metabolism of carbohydrates with exquisite precision. What the body does with sugar can now be reduced to an orderly sequence of chemical processes that include carbon dioxide, pyruvic acid, adenosine diphosphate, etc. This cycle of glycolytic metabolism is exemplary of progress in biochemistry and stands as an achievement of distinction. It is no mark against Hans Adolph Krebs, winner of the Nobel Prize in 1953, that his efforts do not explain or account for or even address just what it is about chocolate that has me reach for seconds. An authoritative understanding of the chemistry of digestion is not the same as—and may not be a step toward—a richer understanding of food preferences or the aesthetics of dining. It may be irrelevant to whether we use shallots or spring onions, balsamic vinegar or fresh lemon juice.

So much for digestion. What is entailed when we claim to have made revolutionary strides in our understanding of *consciousness and mental life*? More is required to effect a revolution here than shouting a rallying cry or singing a victory ode. Presumably, a revolution overturns something. What has long been held regarding consciousness and mental life that must now be abandoned or significantly modified in light of the new "cognitive neuroscience"? Is the revolution one of discovery or retreat? Has progress been won at the price of irrelevance? That is, has more been discovered about the jots and tittles of, say, memory and emotion, but chiefly by limiting the instances of each to what is finally removed from life as lived?

The aim here is not to brandish the tools of the scold. Rather, it is to pause—to catch our collective breath, as it were—and to ask just how far we've come from the more remote sources of these issues and from understandings reached in the centuries leading up to our own. I would be satisfied if readers developed or revived respect for the special nature of mental life and for the challenges it presents to the more complacent forms of "scientism"; a developed or revived respect for a much maligned "folk psychology" that even now, amid the MRIs and gene maps, seems to be the one horse running in a direction the rider recognizes.

ACKNOWLEDGMENTS

I am indebted to my editor at Columbia University Press, Wendy Lochner, for showing an interest in the book from the start and for being supportive throughout the various stages of publication. Robert Fellman had a comforting light hand in editing the manuscript but used it with precision.

The atmosphere within that precious duchy that is "Oxford Philosophy" has been nurturing now for some seventeen years and, as many of my colleagues know, a separate chapter would be needed to record the contributions arising from scores of conversations, attended lectures, questions raised during my own lectures, and the more casual colloquies that punctuate life amid the duties of the term. To cite a special few among the many, I thank Peter Hacker, David Wiggins, Roger Crisp, Anita Avramides, Richard Sorabji, and Richard Swinburne for contributions to my own thoughts on a range of topics bearing on the aims of this book. Readers must not blame them just in case lessons well taught were poorly learned.

To speak of atmosphere is, of course, to thank my bride of some forty years for all that gives ever deeper joy and comfort, day after day, decade by decade.

So, here's another for Ciny . . .

But what consciousness is, we know not; and how it is that anything so remarkable as a state of consciousness comes about as the result of irritating nervous tissue, is just as unaccountable as the appearance of the Djin when Aladdin rubbed his lamp, or as any other ultimate fact of nature.

—Thomas Henry Huxley,
Lessons in Elementary Physiology (1866)

Consciousness . . . is on the point of disappearing altogether. It is the name of a nonentity, and has no right to a place among first principles. Those who still cling to it are clinging to a mere echo, the faint rumor left behind by the disappearing "soul" upon the air of philosophy.

—William James,
"Does Consciousness Exist?" (1904)

CONSCIOUSNESS AND MENTAL LIFE

1

THE GREEKS (AGAIN) AND THE "CONSCIOUSNESS" PROBLEM

Were this intended as an addition to the ambitious offerings of the MIT Press and designed to enlarge its appreciative community of readers, this first chapter might best begin with the neonatal rhesus monkey. Here is a primate still innocent as to the ways of the world but possessing brain cells that respond uniquely to the distress cries of its own species. Presumably thus shaped by evolution, these cells surely must have analogs within the human brain. Why not, then, posit pressures that would shape the brain of *Homo sapiens* in such a way as to foster communal modes of organization dependent, as we now know, on aesthetic values? By the time the opening chapter reached page 4, readers would be urged to give credence to the theory that the best understanding of the summoning power of the adagio movement of Schubert's Quintet in C Major is by way of evolutionary biology. And, if evolution favors "Schubert cells," how much greater must be the shaping of "consciousness cells." Alas, so much for *the problem of consciousness!*

Such neat solutions will be considered in later chapters. At this point, it is important to consider the nature and the source of the problem of consciousness, which more than one contemporary philosopher has declared to be the core problem in philosophy of mind.[1] As such, other problems, such as memory, problem solving, ingenuity, aesthetics, motivation, emotion, etc., are either less problematic or presumably are destined to be settled once consciousness has been dealt with adequately.

As it happens, however, there is nothing in memory, problem solving, originality, or the ordering of objects according to specified standards of beauty that requires consciousness on the part of the entity that accomplishes such tasks. Indeed, as William James observed a century ago, there is very little that we do by way of engaging the challenges of daily life that could not be done without the addition of conscious awareness.[2] Even motives and feelings, if understood as no more than processes or mechanisms by which to induce activity and modulate the "gain" in the system, can be incorporated into explanations requiring no role for consciousness. In light of this, it is far from clear that such complex functions would be explained if the so-called problem of consciousness were settled. Furthermore, once consciousness enters the scene, the already complex phenomena of memory, motivation, and the rest take on a new character radically different from any that would be supposed were the entity a robot.[3] Granting, then, that the problem of consciousness is in some sense the "core problem" in philosophy of mind, it is a problem that adds yet other dimensions to what might seem to be distal to the core.

There is something else that is, as it were, problematic about the problem of consciousness, at least as understood within philosophy. If one relies on contemporary scholarship, there would seem to be a more or less settled position to the effect that the problem is relatively recent in its appearance, largely (and, on some accounts, wisely) overlooked by those very ancients who bequeathed most of the other problems that philosophers would come to claim as their own.[4] Neither Plato nor Aristotle, we are often told, is found dilating on the problem or even recognizing it. Plato, to be sure, was skeptical about perceptual sources of knowledge and defended a theory of the *psyche* (soul, or, more loosely, that which grounds our comprehension of truth) that removes it from the realm of materiality. Clearly, Plato is an early source of that related problem—the "mind-body" problem—but (again, we are told) he cannot be said to have directly addressed the problem of consciousness as such.

Aristotle, of course, as the more biologically inclined commonsense psychologist, took for granted that perception, learning, emotion, etc. occur within the framework of waking life. His theoretical orientation finds him enlarging the functions and powers of perception. It would be a plausible inference to expect Aristotle, faced with what we take to be the "problem" of consciousness, to reply that it is no more or less problematic than perception itself and is best treated as an essential property of much of the animal economy. It has been suggested further that what Aristotle would resist, however, is any version of the problem thought to arise from the alleged duality of mind and body, a point to which I will return.

To say that this characterization of ancient thought is widespread needs to be qualified, for it is surely understood as a controversial claim among those who engage in the systematic study of that thought. In "Why Isn't the Mind-Body Problem Ancient?" Wallace Matson does acknowledge, in the works of both Plato and Aristotle, a recognition of the special nature of the psychic.[5] Nonetheless, he concludes that for both Plato and Aristotle sensation is a natural, corporeal reaction to external events impinging on the sense organs. Thus, having no theory about some inner world of representations accessible only to "mind," neither Plato nor Aristotle needs to wrestle with a mind-body problem. As Matson says: "From Homer to Aristotle, the line between mind and body, when drawn at all, was drawn so as to put the processes of sense perception on the body side. That is one reason why the Greeks had no mind-body problem."[6]

The matter, however, is more complicated than this. Aristotle's analysis is ancient but not dated, and it is useful to consider at the outset, if only to test the interesting claim that his age was oblivious to the finer points at issue. His examination of mind-body relations can be found in a number of his works but, for present purposes, the concentrated version of his analysis is most readily found in book I, section 4 of *On the Soul*.[7] It is in this section that he critically reviews various theories (for example, that the soul is a "harmony" or, following Empedocles, that it is some sort of ratio of constituent elements). He rejects the "harmony" theory chiefly because, if it is intelligible at all, it must refer to relationships among quantities or magnitudes that are localized in space and subject to motion. Such properties, however, must refer to parts of the body and to places therein. But, he says, "there are many and various compoundings of the parts; of what is thought, or the sensitive or the appetitive faculty. . . . And what is the composition which constitutes each of them?"[8]

Just after raising this question, Aristotle notes the tendency to speak of the soul as being pained or fearful or bold. As such states are regarded as modes of movement and change, one might be led to think that the soul, therefore, is thus moved and altered. Granting that such states arise from alterations and motions, this does not warrant the conclusion that the soul is thus affected; rather, the changes originate in the soul, where this term refers not to a place but to a principle of action. Soul, Aristotle notes in his very definition of the word, is a first principle of animate things, in virtue of which a living entity possesses and expresses various powers.

Against those who would have the soul desiring or fearing or being bold, Aristotle offers a schoolmaster's correction, still instructive, especially if the word "brain" or one or another part of the brain is substituted for "soul": "To say that it is the soul which is angry is as if we were to say that it is the soul that weaves or builds houses. It is doubtless better to avoid saying that the soul pities or learns or thinks, and rather to say that it is the man who does this with his soul."[9]

As Aristotle considers the nature of thought (and here "thought" serves as a permissible variant of "consciousness"), he is prepared to accept the substantial difference between it and the merely corporeal properties of the body and its parts. The latter undergo degeneration through injury, disease, and advancing age. Such changes are at the expense of the performance of various parts—eyes, ears, hands. When extensive enough, pathological changes come to impair all the functions that rely on the workings of such parts. Accordingly, when thoughts (or the various aspects of conscious life) are seen to become degraded, it is not the thoughts themselves that have changed but the states and functions on which thoughts depend for their contents and executive functions. Thought as such does not move; it is "impassable." And, as it does not move, it does not *self-move*.[10] For Aristotle, movement and change are the signal features of matter. For matter to rise to the level of *life*, however, the movement and change must be of a special kind. The movement must be of the sort that enters into the processes of nutrition, growth, repair. Wherever plant and animal life is found, there are at least two powers that are required: a *nutritive* power and one or another form of *reproductive* power. Absent the first, individual life is lost; absent the second, the species is lost. Thus, when Aristotle takes "soul" to be the "first principle" of living things, he would have "soul" understood as the constellation of life-sustaining processes. At higher levels of organization, these powers are extended to include locomotion, sensation, and forms of animal "intelligence." With human life there is added that rational power generative of cognitive functions that are taken by Aristotle now as being, in principle, "impassable." This would seem to remove them from the domain of matter-in-motion.

Does all this make Aristotle a "dualist"?[11] This is an anachronistic query not unlike "was Democritus a particle physicist?" Democritus asserted that reality is ultimately decomposable into indivisible particles—α τομος being that which cannot be cut further. His position on atoms was part of an overall cosmology. Aristotle, too, adhered to a cosmological theory according to which the order and lawfulness of nature expressed an originating and designing intelligence itself not reducible to matter in motion. Ever the teleologist,

Aristotle understood the corporeal features of life in terms of ends to be served and realized. Rational choice is toward goals that require perception and locomotion, even memory and imagination, not to mention nutrition and general health. All such modalities may be understood in their respective material and efficient causal roles, but neither together nor separately do they constitute the ultimate goal or aim as such.

Aristotle was as much the biologist as the logician when considering yet other theories that would have soul suffusing the body as a whole, in the form of some sort of subtle matter, as in the atomistic theory of Democritus. But such a theory would require two bodies to be in the same place! The animal can't begin to move until each soul-unit moves in just the right way. If this is to explain the movement of the animal, there is then the task of explaining the causes of the movements of the units themselves, *ad infinitum*.

Defective theories thus eliminated, Aristotle is able to return to the central question of just how it is that modes of thought, feeling, and action are brought about. Considering such powers or states as "knowing, perceiving, opining, and further, desiring, wishing and . . . all other modes of appetition," Aristotle questions whether such discrete functions require a soul divisible into parts, each serving one of these powers. Tested here is a version of today's "modularity" theories. It is challenged by Aristotle on the grounds that such an account leaves unanswered the question as to what it is that holds the parts together. We perceive a dark liquid having a distinct odor and giving off steam. We know it to be strong coffee, about which we have a definite opinion; viz., that it helps the day get started. We desire it and hope that there is more than the cup before us. In this set of functions—perception, recognition, recall, desire, volition—the achievement is a fully integrated ensemble held together as a whole. Were the soul to function as a set of discrete modules, the question that would arise is just how the separate operations are held together to form a complete and seamless psychic occurrence. As the modularity theory cannot explain how the parts of the soul (if it has parts) are integrated, it fails utterly to make sense of the associated claim that the soul holds various parts of the body together as functional systems.[12] All such notions having failed, Aristotle concludes book 2 with the observation that "the question whether the soul and the body are one . . . is as though we were to ask whether the wax and its shape are one; or generally, the matter of a thing and that of which it is the matter."[13]

What is implied by this and similar passages is that "soul" is not a "something" but an operative principle such that what would otherwise be merely dead matter now functions in a characteristic manner. However, such an account is not "on the side of the body," for the same account is required to explain our wishing that there were more coffee to help us get started on the day. There is not a divided assembly of perception + knowledge + wish + desire; there is the complete thought as such.

As for the functions of "soul," it is in book 3 that Aristotle develops the distinctions between perception and thought: between the sensory recording of events and an understanding of them. Considered as recording instruments, the special senses are not prone to the errors that stalk attempts at understanding. Thinking, on the other hand, is associated with opinion and belief. It goes beyond the sensory data and reflects a power or process different from perception. Granting that sensation in ancient Greek thought is, to use Wallace Matson's phrase, kept on the side of the body, *thinking* is another matter entirely. Granting further that there would be no ancient version of the mind-body problem triggered by the facts and assumed processes of sensation, the ancient Greek philosophers would still find in *thinking* (deliberation, understanding, judgment) any number of problems of the mind-body sort. This is evident in Aristotle's comments about intellect or reason (νους), which he declares to be immovable and enduring, properties true of nothing merely material.[14]

It is also in book 3 that Aristotle offers a suggestive analysis of the nature of perception, imagination, and comprehension.[15] The sensory organs supply creatures with representations (*phantasmata*) of objects in the external world. Imagination (*phantasia*) permits the retrieval of this information, and the overall process, as a material transaction between organ and world, carries nothing of "truth" or "falsity." Rather, a series of concrete steps unfolds in much the way a heated stylus produces a copy of its shape in softened wax. But with intellect or understanding there is *belief* and also *conviction* as to the meaning or significance of what is perceived, and it is only by way of human reason that these aspects of experience come about.

These considerations move Aristotle toward a distinctive and distinguishing conception of mind, as such. Noting that, in principle, everything is a potential object of thought, it cannot be the case that the power or faculty of thought has mixtures of material properties. If there were a thought-making device, everything thereby thought would include in some way properties of the device itself. (Crudely put, if thought were produced by a Cuisinart, it would be *chopped*!) A most suggestive conclusion is then reached: "Mind, in order, as Anaxagoras says, to dominate, that is, to know, must be pure from all admixture. . . . It follows that it can have no nature of its own, other than that of having a certain capacity." Thus, that "in the soul which is called thought . . . is, before it thinks, not actually any real thing."[16]

The knowing mind has, as it were, fully taken in the object of knowledge, in a manner that flexibly accommodates all the nuances and particulars. Were the mind a mixture of colors or shapes or odors, then any object absorbed into the cognitive realm would be imbued with such properties, which is patently not the case. Considering only this much, it is clear that Aristotle and, for that matter Anaxagoras, recognized that there is a core problem in philosophy of mind, and Aristotle was able to offer, if not a solution, then at least a clarification of what is problematic about "mind." Any attempt to absorb mind into the realm of material or corporeal *parts* of things will result in implications defeated by the very manner in which the objects of thought are thought, defeated by the very elasticity and capacity of the thinking process. Is the mind, then, some invisible, massless *something*? No, it's not actually any real thing at all! It is not a "process," nor is it a set of "functions." It is the conceptual space within which we find, alas, the objects of thought. Intellect or understanding, as such, is to be understood as a power by which judgments are formed, beliefs are fashioned, convictions aroused. These are all prone to error and deception, whereas the entirely material transactions at the level of sensation are not.

In this same context, Aristotle refers to the soul as "the place of forms," by which he means the powers that convert elementary sensations into coherent wholes. Something of a Gestalt psychology is prefigured in such notions. A bowl of soup has properties that excite the senses of temperature, taste, vision, hearing, and smell. Once the soup is sipped, however, there is not the experience of distinct and separate sensations but a fully integrated experience. It is just a power of ψυχη that achieves the integration, a unifying power associated with thought and perception.[17] The power is basic and thus not reducible to some other power, some other faculty, some other "module."

On the specific question of whether Aristotle was alert to the special nature of consciousness the treatise on the soul leaves little doubt. As Victor Caston has noted, Aristotle regards consciousness as both an intrinsic feature of mental life and as a higher-order process, citing in this connection passages in *On the Soul* where Aristotle notes that we perceive that we are perceiving.[18] This is taken up chiefly in chapter 2 of book 3: "Since we can perceive that we see and hear, it must be either by sight itself, or by some other sense . . . [and] . . . either the process will go on *ad infinitum* or a sense must perceive itself."[19]

If to perceive that we are perceiving we require a separate mental power, then yet another such power would be required allowing us to perceive that we perceive that we are perceiving, *ad infinitum*. The way out of the bind is to accept that perception carries (conscious) awareness with it intrinsically and that perception and thought possess a unifying power over the registration of different and multiple objects and properties that are presented. Noting that "it is impossible to pass judgment on separate objects by separate faculties,"[20] Aristotle adopts the common-sense understanding that thought and experience are unified. If heat were judged by a hot-sensing faculty and cold by a cold-sensing faculty, then neither of these nor both together could generate the judgment "hotter than."

Aristotle's reference to the fact that we perceive *that we see and hear* clearly points not to a theory of consciousness and self-consciousness but to an everyday fact of experience. Later commentators on this passage were at pains to find its deeper significance, as has been recently analyzed by J. Noel Hubler.[21] Finding the wide variations in these commentaries, Hubler concludes: "The attempts by so many serious and reflective ancient Philosophers to understand self-awareness . . . should serve to dampen the hopes and pretensions of modern theorists to take self-awareness as a primitive, privileged, and unassailable basis for understanding the mind."[22]

But he follows this very conclusion with the useful interpretation of the main mission of the commentators; viz., anchoring whatever their conception of self-awareness might be to their more fundamental interest in the nature of *truth* as such. On this point, Hubler's analysis is unexceptionable, but it leads to a conclusion strikingly different from the one he reaches. If the commentators do not provide what Hubler calls a "transparent" theory of self-awareness, it is because self-awareness itself is "transparent" and requires no theory. Granting without further (and pointless) analysis that persons have awareness and self-awareness, interesting questions then arise as to whether awareness at the level of perception or awareness solely at the level of reason or intellect is trustworthy. For the present purposes, however, it is sufficient to recognize that, if consciousness is problematic, the problem did not begin with Descartes.

In this same connection, it should be recalled that perception has wide-ranging powers in Aristotle's psychology. He argues, for example, that "thinking, both speculative and practical, is regarded as a form of perception; for in both cases the soul judges and has cognizance of something which is."[23] This much acknowledged, the question again arises as to whether in the end Aristotle is committed to a conception of mental life exhausted by the corporeal and biological constituents and principles of life itself. Put in stark terms: *was Aristotle a materialist*? It should be clear by now that the answer is no, but support for that answer calls for journeys to and through metaphysical and logical works far afield from the current issues.

There is a useful shortcut, however, in books 11 and 12 of his *Metaphysics*. Here Aristotle addresses that most fundamental question, which is expressed in modern fashion as: *Is physics complete*?[24] This is a question to which I will return frequently in these pages. If, indeed, there is no feature of reality at once really existing and also outside the realm and range of the physical, then the ultimate science will be natural science (*phusis*). "If, then, material substances are the first of existing things, physics will be the first of the sciences; but if there is some other nature and substance which exists separately and is immovable, then the science which treats of it must be different from and prior to physics, and universal because of its priority."[25] The passage is relevant, apart from its ontological claim, because it appears within a general discussion of sciences falling within the sphere of human understanding or wisdom (σοφια) and therefore compatible with modes of understanding hosted by our rational, mental life. We are able to explore such possibilities because rational thought itself is not constrained by the body's parts or processes, by the materiality of life.

It is in book 12 that implications are drawn out as regards the divine "unmoved mover" and the relation between that divine thought and human thoughts and desires. The central argument is toward the conclusion that the "first science" is theology, for it is about first things—about being *qua* being rather than particulars and their changeable features. Thus, if the movement of anything is impelled, then, whatever it is, it could be otherwise, this being the sense in which any of the mere facts of the natural world could be different from what they are. The unmoved mover, however, could not be otherwise, for its movement is independent of any source other than itself. It is this absolutely necessary first principle that is "the first principle upon which depend the sensible universe and the world of nature. And its life is like the best which we temporarily enjoy."[26] We have this sort of life only for that brief time (*mikron chronon*) when deeply contemplative human thought is concerned with "that which in the highest sense is best. . . . And thought thinks itself through participation in the object of thought . . . so that thought and the object of thought are the same."[27] In this connection, it cannot be emphasized enough that, for Aristotle, though there is no radical dualism of soul and body—*psyche* and *soma*—there are ineliminable differences between *mind* and body. Aristotle's conception of *psyche* matches up with principles of organization and empowerment. It is the "first principle of living things" manifesting itself in the nutritive, reproductive, locomotor, sensitive, intellective, and rational powers as these are witnessed among one or another class of living things.

Sensation (*aesthesis*), on this account, is just that psychic power (*dunamis*) in virtue of which a creature is affected by stimuli such that the things and items in the external world come to be known. Here everything is on that side of the line where one finds the somatic, but this barely scratches the surface of Aristotle's moral and cognitive psychology. No aspect of the rational comprehension of the world, no aspect of that scientific knowledge of ultimate or "final causes," is explicable in terms of corporeal parts and processes. The intellect can be affected by states of the body; for example, under conditions of excessive excitement, it will fail to make proper use of the φαντασματα provided by the sensory organs. But intellect as such, at least in its episodic engagement of "first things," operates independently of the body and is not reducible in its operations to those of the body either in part or as a whole.[28]

Enough of Aristotle. Far less space is needed to establish Plato as a philosopher fully aware of the mind-body problem and as one who addressed it repeatedly, if not always in a manner that would appeal to contemporary sensibilities. Referring to Plato's dialogues, T. M. Robinson is surely correct when he says that "they contain the first fully articulated account of the relationship between soul (*psyche*) and body (*soma*) in Western literature."[29] This is an estimation that Wallace Matson would share, so it is again necessary to consider the version of the mind-body problem that would distinguish Plato's understanding from what is current in the philosophical literature.

Brevity here is hazardous, but, with due qualification, I would be inclined to say that the current philosophical literature understands the mind-body problem as at once an ontological problem and (relatedly) an algorithmic problem. The ontological problem harkens to the "physics is complete" maxim such that nonphysicality is rejected *a fortiori*. The algorithmic problem arises owing to the very phenomenology of mental life, a phenomenology that seems to have little or nothing "physical" about it. How is it, then, that a purely physical system operating on physically defined inputs somehow represents whole clusters of seemingly aesthetic, moral, and social domains also having little or nothing "physical" about them?

If this is at all a faithful summary of current understandings of a version of the mind-body problem, then Plato's position is very nearly contemporary. In several of the dialogues, especially the *Phaedo*, the *Republic*, and the *Laws*, there are developed arguments against the plausibility of absorbing the psychosocial and rational aspects of life into the domain of bodily functions. The ontological mismatch is obvious to Plato, as much so in the consoling passages of the *Phaedo* regarding the soul's afterlife as in the *Laws*, where Cleinias and the Athenian directly address the question of how soul *qua* soul can control a physical body. "Soul" in the Platonic system is the source of all movement, including that of celestial bodies. Orderly motion is the product of the "good" soul, wild and frantic motion of one that is less good, or evil.

Then, directly, the Athenian asks: "Of what nature is the movement of mind?—To this question it is not easy to give an intelligent answer."[30] All can witness the movement of the sun across the sky, but no one can see the soul that is the source of the motion. And to try to see it by looking at the sun is to suffer an induced blindness, "making ourselves darkness at midday."[31] Rather, one must consider the matter obliquely, as it were, in terms of the various modes of motion, finally selecting the type that fits the mind's mode. The conclusion offered by the Athenian is that "mind" moves in a manner that is ordered, "circular," proportioned, "like the motion of a globe," all this providing no more than an "invented . . . image."[32] This is a thesis that will be considered below, in chapter 7, in connection with Descartes' own approach to the question of mental causation and the sense in which one might speak of the mind's "motion." When the Athenian analogizes motion of the soul to that of a globe, there seems to be an anticipation of Descartes' own rejection of any fixed starting point for mental life. Instead, it is to be understood as self-sustained and ongoing. But further elaboration of this point must be deferred for now.

We see once again that the correct position on the matter of ancient thought is that Aristotle and Plato both confronted the mind-body problem in different ways but, to some extent, in a manner similar to contemporary encounters. Perhaps, then, the better answer to Wallace Matson's summoning question is not that the ancient Greek philosophers just missed seeing the problem, but that, never assuming that "physics is complete," they found in it no threat to the larger metaphysical framework within which their inquiries proceeded. The point to be emphasized here is that metaphysical frameworks, unlike physical frameworks, are not constrained or verified by observation and measurement. That Plato and Aristotle did not consider mind-body issues within the framework of dogmatic physicalism does not render their position dated or simplistic. Rather, it stands as a challenge to dogmatic physicalism and offers as the very terms of the challenge the seemingly nonphysical properties of mental life.

That the inquiries of Plato and Aristotle now seem tentative and framed in terms of metaphor or simile is evidence not of their opacity but of their respect for the limits of metaphysical and scientific analysis in matters of this sort. As will be explored further in the next chapter, the very modes of philosophical inquiry in the Hellenic world differed from today's, where linguistic analysis continues to be the method of choice when not an end in itself. These differences duly noted, the contemporary student might well ask whether current modes of inquiry have succeeded where Aristotle and Plato failed or have shown that the ancient understanding was based on the pardonable ignorance of an earlier time. The space devoted to Aristotle and Plato in this chapter seems warranted, for the progress, such as it has been, since their flourishing can be assessed only to the extent that we are aware of just where they left matters. Let the reader decide.

If the decision is to be adequately informed, however, at least these main points, which will be repeated and developed in succeeding chapters, should be kept in mind: First, what makes something problematic, either in science or in philosophy, is its resistance to absorption into currently accepted ontological frameworks and explanatory schemes. "Consciousness" is a problem, then, to the extent that there is widespread agreement that reality is exhaustively constituted by physicality. Unless consciousness is "code" for what is finally physical, it is a renegade concept. Next, just how a problem is to be settled also becomes more or less "official" in various epochs: revelation, intuition, observation, argument, consensus, authority—all sorts of debate-stopping conventions have been adopted by the best and the brightest over the eons. We (collectively) have come to regard linguistic and conceptual analysis as the right way to work through our own metaphysical conundrums. As a result, we now have numerous philosophies of mind, all dealing with greater or lesser agility with all sorts of problems, even as the central problem—that of consciousness and its unique contents—remains fully intact. Then, too, dread "Cartesianism" has been given such widespread publicity in books and journals that there is a worrisome avoidance of just what Descartes actually had to say, why he thought he had to say it, what his critics said in rebuttal, how he handled the criticism, and how original he was in his treatment. In that rich colloquy, much of what now passes for leading-edge thought proves to be reheated—and often stale—by comparison. Finally, granting that the foundational content for the physical sciences is matter and energy, it is important to keep in mind the question of the foundational content for "cognitive neuroscience." I shall argue that this content just is "folk psychology," the authority of which is rejected at the cost of cognitive neuroscience having a subject.

2

THE PROBLEM OF CONSCIOUSNESS "SOLVED"

Every year, a spate of philosophical articles and books offers a solution to the problem of consciousness, or a more scientific explanation of it, or the breaking news that there is no such thing, though the remnants of village superstition continue to convince us otherwise. Widely cited in this connection was Daniel Dennett's *Consciousness Explained*.[1] Given the provocative title, hopes were high as the pages were thumbed but, in the end, notwithstanding to the contrary the author's sprightly discourse and technical competence, what was delivered was less an explanation than a metaphor—and not a very convincing one. Turning to the brain, Dennett made much of the proposition that consciousness is not here or there within the cranium, thus putting to rest something called the "Cartesian Theater." The theater in question is one that depicts on some sort of screen or stage the contents of consciousness. (As will be noted in the next two chapters, Descartes was specific in rejecting such an idea, but "anti-Cartesianism" is now something of a professional sport in philosophy and there is no reason to single out Dennett for rebuke on this point.) Offering as an up-to-date corrective what is really the traditional view (i.e., that such global features as consciousness are not localized), Dennett argues instead that the brain functions as a collection of "semi-independent agencies."[2] The different functional units within the collective contribute a portion to the whole.

Much more will be said later in the chapter on the positions advanced by Dennett, for these have been influential and repay close attention. Here I would note only that nothing in *Consciousness Explained* actually explains consciousness, for nothing in the book actually addresses the phenomenology of consciousness itself. It is more like a book on how to prepare sauces for various preparations without indicating that the ingredients, properly combined, will produce gustatory sensations of the desired type, or like a treatise on optics that somehow never gets around to the fact that we actually *see* things. What is omitted is finally what is essential about consciousness. And what is essential about it turns out to be essential about us, too. A few words are in order, then, on the matter of *essentialism* and whether it is guilty of ontological profligacy.

There is a strong cognitive tendency, as evident in ancient as in modern thought, to *essentialize* as we enter into commerce with the objects of the world. We take apples and rabbits and baseball bats to have "essential" properties such that we recognize these items to be *the kinds of things they are*, even though they all come in different sizes and colors. Systematic and scientific forms of essentializing appear in such special fields as taxonomy and chemistry. The periodic table is a veritable codebook of "essential" properties: copper, for example, is what it is and nothing else.

Without pausing (until the next chapter) to consider whether Descartes was on the right track in declaring himself to be "essentially" a *res cogitans*, we can surely agree that our state of conscious awareness is a feature that trumps all others in the matter of epistemic authority. Each of us has the last word on whether we are in such a state of awareness (consciousness), though we might always be wrong in classifying or "essentializing" anything that comes into the range of that awareness. As dualisms go, then, the mind-matter variety is one of a number of plausible schemes. There is surely nothing weird or demonstrably false about declaring reality to host two *essentially* different kinds of things: those that are mental and those that are physical.

Another plausible scheme might be this one: All really existing items can be classified as possessing or lacking consciousness, and some that possess it may lose it either episodically or finally. Moreover, those that lack it might come to possess it. Furthermore, that an item is physical or mental does not sufficiently or necessarily establish whether it possesses or lacks consciousness. Now, on this scheme, there is no more a "problem" of consciousness than there is a "problem" of bivalent atoms; it is simply one of the constituents of reality. That is, it is not problematic in and of itself. Rather, it becomes problematic only on the further assumptions that (a) all really existing items are physical and (b) mental properties as such either do not exist at all or, if they exist, they exist dependently on physical properties.

Aristotle, Plato, and their age surely accepted that part of the common-sense "essentialist" scheme that includes consciousness as an exceptionless feature of the community of mid-size and large animals. Its episodic character arises from the need for sleep, and its loss arises from serious illness, trauma, or death. On this view, there is no more a problem of consciousness than there is a problem of sleep. Might a table be conscious? No. Why not? *Not* because tables are physical entities, for human beings are also physical entities. Rather, tables are simply not the *kinds* of things that possess consciousness or, for that matter, feathers. Why should this be any more problematic than any other feature that renders anything the kind of thing it is?

If consciousness and mental life are problematic in respects different from those recognized by ancient and later philosophers, and if a culprit is to be elected by vote, most ballots would contain the name Descartes. I should say straightaway that this is not entirely fair but, again, the main defense must be deferred to the next chapter.

Assuming that it is the mere word *consciousness* that has caused such trouble, it was actually Locke who used it in an especially controversial way, and it is useful to discuss this at least briefly.

Difficulties appeared early when Locke's contemporary translators looked for words that matched the meaning Locke had given the word. Indeed, "consciousness" in the Lockean sense would not find its way into the Chinese vocabulary until the nineteenth century.[3] To the extent that it was used before Locke's influential *Essay* it was based on the Latin roots—*cum scio*—which convey joint or shared knowledge, understandings that obtain within a social community. The *Oxford English Dictionary* offers as the earliest acceptation of the term "Knowing, or sharing the knowledge of anything, together with another; privy to anything with another" and supplies the relevant Latin phrase *alicui alicujus rei conscius*. Hobbes understood the word in this older social sense of shared knowledge: "Where two, or more men, know of one and the same fact, they are said to be Conscious of it one to another."[4]

Used instead to refer to something like a private domain accessed through the introspective resources of a single person, the word takes on a different meaning, thanks here not only to the culprit, Descartes, but perhaps especially to Locke. What is it, then, that pegs Descartes for such sustained criticism in the matter of current quandaries and correctives regarding consciousness and mental life? Gilbert Ryle, in his pathbreaking book *The Concept of Mind*, presents a summary of what Ryle understood Descartes' thesis to affirm: "Human bodies are in space and are governed by the mechanical laws which govern all other bodies in space. Bodily processes and states can be inspected by external observers. . . . But minds are not in space, nor are their operations subject to mechanical laws. . . . Only I can take cognizance of the states and processes of my own mind."[5] This is not a wayward abbreviation of Descartes' position, though as with any summary of a complex argument it leaves much unsaid. At a common-sense level, it is quite obvious that the reader has no idea of what I plan to write in the next chapter, though, as it happens, I have a very clear idea of it. Similarly, I have no knowledge whatever of what the reader is planning for supper tomorrow night, but the reader may already have prepared the menu. There is surely nothing startling about the notion that what is available by way of first-person accounts may not be otherwise available.

I leave unexamined "processes" and "states" that figure so centrally in contemporary cognitive (neuro)science, though in later pages I will draw attention to the jejune status of these terms when intended as explanations. Here it is sufficient to call attention to the fact that such terms tend to carry their own explanations with them, leaving the skeptic in a kind of lexical wilderness. As the skeptic scans the flora and fauna of neurophysiology, it is far from obvious what either a "state" or a "process" is or what the means might be by which to identify either or both. Electrical circuits tend to have binary states—ON and OFF—whereas other systems display continuous variation in, for example, temperature, velocity, or consumption of power. The nervous system is, of course, continuously active and, as such, may be said to be ON, but this is too molar a description to be of much explanatory value. Processes, too, can be identified with precision: the Krebs cycle summarizes the sequence of metabolic steps resulting in the body's use of carbohydrates. But, again, the brain is a veritable labyrinth of "processes," none of which seems anything like the mental "processes" presumed to depend on them.

Staying with Ryle, it is clear that the claim that "mental" states and processes are subject to mechanical laws is distinct from the claimed uniqueness of first-person reports of thoughts or feelings. Let us stipulate that the next generation of "intelligent" automobiles will provide a device for thwarting thieves. To unlock the door and to activate the ignition, it is necessary for the driver's fingerprint to be "read" by the device. All this may well be subject to the mechanical laws that govern bodies, but the owner of the vehicle nonetheless has privileged access to the automobile and its various functions. Thus, it is not the issue of "privileged access" *per se* that is problematic—at least it is not solely this that is problematic—but the additional (though unnecessary) stipulation that physical laws are not determinative in mental life. So if Descartes is a culprit, he is so for defending a psychophysical *dualism* long out of favor within the community of those committed to the proposition that, once again, *physics is complete*.[6]

It is worth considering this proposition again, for no less than Carl Hempel, for all his commitment to scientific modes of explanation, found it problematic. Hempel knew that at any stage in the history and development of the physical sciences there were positions firmly taken that would ultimately have to be abandoned. Thus at no stage up to the present could it be factually asserted that "physics is complete." If, instead, the proposition is no more than promissory, it is simply a will-o-the-wisp and not of any philosophical value.[7]

Defenses of physicalism often proceed from criticisms of any and all alternatives. To the extent that such alternatives call for nonphysical modes of causation, the critic need only expose the (alleged) incoherence of a nonphysical cause of physical events. We see a version of this in David Papineau's defense of physicalism. Referring to the putative "mental" causes proposed as additions to merely physical causality, Papineau writes: "If there were such forces, they could be expected to display some manifestation of their presence. But detailed physiological investigation failed to uncover anything except familiar physical forces [i.e., forces that are not *sui generis* mental]. In this way, the argument from physiology can be viewed as clinching the case for completeness."[8]

Here, then, is an extension of the "physics-is-complete" thesis, now including that *physiology is complete*. However, the very question of the "completeness" of physics expresses a transparently cognitive disposition that, itself, must be subsumed under the phenomenology of consciousness. The completeness affirmed can only refer to the degree to which a particular model or paradigm of explanation has been accepted as final, total, underived, fundamental. Surely, it is not the case that "physics is complete" in accounting for the classic revival in the architecture of the eighteenth century, thus displacing the more ornamental *baroque*; or Shakespeare's success in having Dogberry refer to comparisons as *odorous*; or the basis on which the decision in *Chisolm v. Georgia* resulted in the Eleventh Amendment. Physics, if it is ever "complete," completely accounts for and predicts the sorts of physical events and properties for which physicalistic modes of inquiry and measurement are authoritative. Just how good the resulting explanations are cannot be settled by physicalistic modes of inquiry and measurement, however, but must finally be *judged* according to an essentially cognitive mode of evaluation and comprehension. In a word, to rule out dualism on the grounds that it is inconsistent with the putative completeness of physics reveals a strain of unintended irony. The sentence itself—*Physics is complete*—is so utterly beholden to considerations of a cognitive and even an aesthetic nature as to be nothing short of paradoxical.[9]

This is not to say that in some still and utterly elusive way a thorough mathematical account of reality might not include an account of whatever it is that yields "consciousness" at the phenomenological level. The equations of quantum mechanics provide what physicists take to be a complete description of the physical world as known. Michael Lockwood has offered a searching and fully informed account of the manner in which consciousness itself may be included within and reveal yet a fuller comprehension of quantum mechanics itself.[10] At the center of notions of "quantum consciousness" is the liberation of the physical from primitive notions of "matter" as such. To the extent that our current descriptions of brain function are tied to the older materialist-mechanist concepts of the older macrophysics, it is not surprising that there is no apparent similarity between our phenomenological reality and thoughtless extended bodies. The quantum world is different, possessing enough that is "mysterious" to begin to look like a candidate for the provenance of consciousness. The hypothesis is not new. William James speculated in his *Principles of Psychology* that, were it not for mental states, the hair-trigger instability of the brain would descend to chaos. In his characteristically jaunty way, he says:

> A high brain may do many things, and may do each of them at a very slight hint. But its hair-trigger organization makes of it a happy-go-lucky, hit-or-miss affair. It is as likely to do the crazy as the sane thing at any given moment. A low brain does few things, and in doing them perfectly forfeits all other use. The performances of a high brain are like dice thrown forever on a table. Unless they be loaded, what chance is there that the highest number will turn up oftener than the lowest? All this is said of the brain as a physical machine pure and simple. *Can consciousness increase its efficiency by loading its dice?* Such is the problem.[11]

This is surely not the place or the time for oracular pronouncements as to whether or how advances in quantum theory will comprehend the problem of consciousness in different terms. If, for example, a Bose-Einstein condensate is an apt model of the manner in which subatomic states in the brain achieve a wholeness or unity that grounds the unity of conscious experience, so be it. If, for example, owing to the fact that brains are composed of precisely what goes into the formation of everything and anything, such that, at the quantum level, "consciousness" is just a feature of quantum states, then we have a defense of *panpsychism*, and so be it, too. None of this, however, would "solve" the problem of consciousness in its phenomenological manifestation, which, in the end, is the only manifestation that is problematic. None of this would, in the end, remove quantum mechanics itself from a *cognized* reality bearing the stamp of a rational mind's reflections on what falls within the range of its competence.

With respect to the putative "problem" of mental causation, there is surely one sense in which ascribing causal power to mental states is unarguably correct. To say that my desire to participate in the New Year's Eve celebration caused me to lift my glass is to speak in a manner that makes sense to any who would wish to explain the behavior of grasping a glass and holding it high. There are criticisms that may be mounted against such explanations,[12] but they boil down to adherence to the very physicalism to which mental causation is an enduring challenge. Except in instances of neurological and psychiatric disturbances, a person's actions proceed coherently and intelligibly from desires and beliefs. That similar or even identical patterns of behavior could be elicited by way of direct stimulation of the brain establishes that some activities are not the actor's *own* in that authentic sense of realizing and satisfying the actor's aims and desires. What makes an action authentic, then, is its connection to a mental event—a conscious desire or belief—uniquely possessed. One reply to the question, "How can such an event *cause* the body to move?" is just another question: "How can purely material activity in neural networks *cause* a conscious belief?" If the first question is designed to expose a mystery, so also will the second.

But here I get ahead of the story. As it happens, "solving" the problem of consciousness is important (presumably) for reasons that go beyond consciousness per se. There are several venerable issues in philosophy that seem so closely associated with consciousness that one might expect them to be settled by the right theory or account of consciousness. Included among these are self-consciousness, the seeming unity of consciousness, and all the first-person accounts that enter into the evidence for emotion, desire, motivation, goals, moral autonomy. At least at first blush it would seem strange to speak of "unconscious emotions and desires," though psychoanalytic theory makes much of "unconscious motivation." But even on the Freudian account, although actors may be deceived as to what impels or guides their actions, they are nonetheless aware of what they take to be the plans and purposes their actions seek to realize.

As for the putative unity of consciousness, there are all varieties of exceptions in the thick casebook of clinical neurology, but for most persons during the waking hours of life, the domain of which they are conscious falls within a unified field of consciousness.[13] Two persons may be conscious of the same apple, but one person does not have two consciousnesses of the apple. Note also that in referring to waking hours, I don't suggest that consciousness is absent in sleep. Dreamers later report vivid experiences, recalled in much the way that waking experiences are. These same later reports leave no doubt but that, during the dream states, the dreamer is as "conscious" of the dream events as would be the case in wakefulness.[14] The reality of these episodes is often so undeniable that, on awakening, the person may remain alert and active *in the dream world* for some minutes. What the dreamer lacks when aware of the dream content seems to be not consciousness but epistemic justification. Were the dreamer to speak, reporting the events and persons encountered in the dream, the reports would form no public or transpersonal record, no evidence to support the reports, and no other with whom to claim a shared experience. Perhaps in the end it is an abuse of language to speak of one as consciously dreaming, but the facts are what they are. The point I wish to make here is just that if there is, in this sense, some sort of dream-state consciousness, then it is unclear that a fuller or more compelling theory of consciousness in the ordinary sense would thereupon account for it.

Distinct from these considerations is a pair of more general questions: First, in what sense is consciousness a "problem," or, put another way, in relation to what other aspiration or project do the phenomena of consciousness create difficulties? Having briefly considered Aristotle's analysis, it is clear that the problem arises chiefly when we attempt to identify "mind" with material bodies, or insist that it is an immaterial "it." If consciousness were just another fact in a world of facts—if it were nothing but that "mind" that Aristotle reduced to a capacity—it would be no more problematic than, say, copper or Shetland ponies. Few refer gravely to "the problem of copper." Consciousness, if it is problematic, must be so owing to the kind of fact it is: a special kind, present to all, but only, as it were, to "one at a time." This, of course, was not a problem for Aristotle, for it is not problematic that perceptual events are individuated at the level of the percipient. It would be radically more problematic were they not!

The second question has to do with the form the desired solution should take. Mathematical problems result in numerical solutions. Mechanical problems are relieved by the assembly of mechanical instruments and devices. Hormonal problems are addressed pharmaceutically. We may ask, then, what *mode* of solution is intended for the problem of consciousness? The temptation, at least within philosophy, is to insist on some sort of linguistic or "conceptual analysis," but one must be careful before investing too much in this. There is an utterly common-sense and widespread understanding of what it means to be conscious of something. If this common-sense understanding is not to be trusted, one must then consider whether a very different concept, once "analyzed," will make contact with the very "consciousness" that was so problematic at the outset. Too often, the analysis that works so well in the armchair proves to be less than serviceable in the marketplace. Later in this chapter I will consider the assets and liabilities of linguistic analysis in dealing with the problem of consciousness. What I hope to make clear is how the adoption of a philosophical method determines just what counts as a problem.

Some philosophers have proposed a radical way out of the bind; namely, by taking the so-called problem of consciousness as akin to the problem of unicorns—the problem of finding a method capable of explaining what isn't there to be explained. For these philosophers, the *eliminativists*, the entire project of solving the problem of consciousness should be abandoned as a vestige of, yes, Descartes' ontological blunder: that of conferring a separate and distinct ontological standing on the material body (the *res extensa*) and on the allegedly unextended mind, the *res cogitans*. What Descartes simply missed, critics say, is the obvious fact that all so-called mental operations merely express the functions of the body and more particularly the brain.

As fresh as this criticism still seems to many, the position was fairly explicitly developed in the eighteenth century by La Mettrie and was at least strongly hinted at by Descartes' own contemporaries, especially Gassendi and Hobbes.[15] It is also peculiar as a criticism in light of Descartes' own very strong inclination to account for nearly all psychological functions (perception, directed movement, emotion, motivation) in uncompromisingly physiological terms, a point discussed further in the next chapter.[16] All this duly noted, more current formulations of the criticism need to be considered, just in case they include something Hobbes, Gassendi, and Descartes himself overlooked.

In a more modern idiom, consciousness as such is acknowledged but is reduced to nothing but a state—ultimately a physical state—of a complex system whose fundamental components, though unlike consciousness, are nonetheless the source of it. In this class of solutions to the problem of consciousness we find one or another version of *emergentism*, of *epiphenomenalism*, and of *supervenience*. The positions are not interchangeable, but there is a common thread running through them.

Emergentism subsumes a variety of explanations sharing the judgment that the phenomena of consciousness emerge from the extremely complex operations of the brain. It is a venerable position, its advocates including Gassendi among Descartes' contemporaries, most of the materialists among the Enlightenment *philosophes* (e.g., La Mettrie, Holbach, d'Alembert, and Cabanis), and, in the nineteenth century, Spencer, Bain, and then the entire school of Darwinians. Evolutionary theory included an evolutionary psychology according to which the complexity of mental life was bound to the evolving complexity of biological life. By the turn of the century, leaders in comparative psychology were successful in demystifying mental phenomena by establishing their analogs up and down the phylogenetic series. Well known to students of the history of psychology, C. Lloyd Morgan advocated Ockhamist restraint in attempts to explain psychological processes. "Morgan's Canon" requires of an explanation nothing beyond what is needed to cover the observed facts. If the best account of the conditions under which spiders weave webs and birds build nests plausibly reaches human architectural initiatives, there is no justification for going beyond the principles of evolutionary psychology in framing a scientific explanation.[17]

Of the philosophers attaching themselves to this broad perspective, one in particular developed the *emergentist* implications in a nearly contemporary fashion. I refer to Samuel Alexander, whose *Moral Order and Progress: An Analysis of Ethical Conceptions* (1889) was explicitly evolutionary in its explanation of the origins and purposes of the moral order. Alexander would develop strong interests in psychology and would come under the direct influence of Morgan, with whom he then had a life-long friendship.[18] His most celebrated work was *Space, Time, and Deity*,[19] in which he defends a version of *emergentism* that might well appear in a current treatise on the mind-body problem: "a process with the distinctive quality of mind or consciousness is in the same place and time with a neural process, that is, with a highly differentiated and complex process of our living body. . . . There is but one process which, being of a specific complexity, has the quality of consciousness." He then concludes that at "a certain level of development," the neural process just has the property of consciousness. The latter emerges from the former as one of the bequests of evolutionary development.

The correspondence between *emergentism* and what has come to be called *supervenience theory* (ST) is so close that a major advocate of the latter, Jaegwon Kim, is inclined to treat ST as a variant of *emergentism*.[20] Kim distills the theory to four core assumptions: (1) All real particulars are physical; (2) Mental properties are not reducible to physical properties; (3) Mental properties are realized by physical properties; (4) Mental properties are real. I shall characterize these four assumptions as follows:

1 = physicalism
2 = essentialism
3 = causal efficacy
4 = emergentism

Employing this scheme, we will be in a position in the next chapter to assess what is gained over "Cartesianism" and, in the process, come to grips with just what Descartes had proposed, but this will be reserved to a later discussion. At this point, I propose to make the now controversial claim that nothing actually "emerges" from something that it is not. And, as room for confusion is wide on such a matter, some degree of repetition must be suffered.

The textbook example of emergent properties is that of water. To wit: Neither hydrogen as such nor oxygen as such has the properties of water. Accordingly, a complete physical account of all that there is to know about these two elements would not permit one to predict that, in the right combination, a clear and potable solvent would emerge. Nonetheless, water is nothing but the combination of two atoms of hydrogen and one of oxygen. Clearly, then, this watery sustainer of life, though having no property in common with either hydrogen or oxygen atoms, emerges from nothing more than a specific combination of the two.

Note, however, that a basic question is begged by such accounts, for it is surely arguable that "all that there is to know" about oxygen and hydrogen presumably would include within the vast store of knowledge that, in the right combination, the two will yield water. Whatever it is that is responsible for the transparency of water is, one would presume, finally explicable in terms of optical transmissivity at the molecular level. That water dissolves other bonds is also to be explained in terms of bonding forces and the physics associated with breaking them. In these respects, being "water" is not unlike being "dressed." When the frantic chap in the steaming shower appears an hour later in a tuxedo with the properly attached carnation, we might be inclined to say something like, "You wouldn't believe what emerged from the shower!" But there is nothing in the later edition that was not present in the initial ingredients.

The key passage in the emergentist script is "knowing all there is to know about. . . . " Transporting the account into the more difficult subject of consciousness, we would have something like this as a candidate explanation:

1. Consciousness is nothing like the electrochemical processes of neural cells.

2. In the right number and interactions of these cells, consciousness emerges.

3. No extent of knowledge about the neural material would predict the properties of consciousness, but these properties, though real, are nonetheless nothing but the product of the complex neural interactions.

Again, the "no extent of knowledge" provision is ambiguous. Certainly complete knowledge of the properties of sodium and of chlorine would include the prediction that they will bind to form NaCl, and that the result of this binding will be a salt soluble in water. Why, then, is it surprising that the properties emerging from the right mixture of hydrogen and oxygen are "watery"? Note how the plot thickens, however, when water is the emergent property. Water's viscosity is, presumably (*pace* Berkeley), independent of the perceptual resources of an observer, but *wetness* is not. If it is the special property of the wetness of water that "emerges" from mixtures of two parts hydrogen and one part oxygen, then one has every right to say that what has emerged is itself not a physical property but a *quale*, and that such *qualia* are themselves part of the problem of consciousness and not part of the solution.[21]

Perhaps the example is poorly chosen. Let's turn to ST as it tends toward the same conclusions reached by emergentists. My computer rests on an oak table. The oak table is *real*. It is not a figment of my imagination and, were the entire kingdom of percipients to disappear at once, there are sound reasons to believe that the oak table would retain all the properties that now compose it. However, the oak table, as a table, has a dependent relation to oak, as oak has to a particular carbon-based molecular configuration. The hardness of the table is dependent on the density or packing of the constituent molecules. In a word, the table *supervenes* on lower-level realities, including ultimately the particle-physics reality that grounds all particulars.

The account here is based on a *layered* versus a *bifurcating* conception of reality, to borrow Kim's useful distinction.[22] There is no bifurcation of reality into physical particulars and ontologically distinct nonphysical particulars, no bifurcation into matter and spirit, bodies and minds. Instead, there are layers of reality, each particular within a given layer as real as any found at another layer. Napoleon really did lose a pivotal battle in Belgium and Cicero really did make the case against Cataline. But the reality of Napoleon's Waterloo, like the reality of Cicero's three orations, is dependent on a lower-level reality of complex biosocial interactions that, in turn, *supervene* on a still more basic level of biogenetic interactions, all the way down to that *terminus ad quem* where one finds the ultimate "particulars," whether quarks or something else.

One might be willing to accept that the table as such "supervenes" on molecular constituents that compose oak as a type and that the latter "supervenes" on more basic carbonaceous elements, and these then on more elementary particles, and these then on subparticles. But in this account, there would be no way of distinguishing oak tables from everything else physical! All that the supervenience theory or model provides in such cases is the (ultimate) sense in which an object is, among other considerations, a physical thing. As the subatomic account in and of itself will not reveal that the thing under consideration happens to be an oak table, so the same account would not disclose that the entity under consideration happens to be conscious. And if there is a serious scientific question as to just how it is that, given the particulate nature of reality, oak tables come about, it is surely a serious scientific question as to just how it is that, given the particulate nature of brains, consciousness comes about. Nor is this the final step. Rather, we would then have to press on with what must be the far more demanding task of establishing how the quite peculiar property of consciousness emerges from the rather prosaic electrochemistry of neural tissue. Added to this is the related task of explaining how the interests, expectations, motives, and beliefs of the person result in consciousness being directed and deployed less as a "state" than as a tool. Not much by way of skepticism is needed to be wary of the entire line of reasoning supporting ST and emergentism in accounting for these ubiquitous aspects of consciousness.

It is entirely unclear as to what is meant by the claim that although consciousness does not match the properties of the constituents of which it is formed, it is no more than a compound of them. At a common-sense level, we would be inclined to say of something having the salient properties A, B, and C that it could not arise from a combination of other things having not one of these properties, though all the while admitting of epistemic limitations on our part; that is, there may be any number of real properties beyond the ambit of our intellectual, perceptual, or moral resources. But in just the sense that we can't make a silk purse out of a sow's ear, we have the strongest resistance to the notion that we can get Napoleon's Waterloo from what, at base, is quarklike. Where an account is utterly promissory, it is not unfair to apply the test of credibility and to begin to think of a reality that really does *bifurcate*.

What, then, of the *epiphenomenalism* advanced in the nineteenth century by Thomas Henry Huxley? Still among the more engaging versions, here is how Huxley put it in "On the Hypothesis That Animals Are Automata" (1874):[23]

> The hypothesis that brutes are conscious automata is perfectly consistent with any view that may be held respecting the often discussed and curious question whether they have souls or not; and, if they have souls, whether those souls are immortal or not. . . . If the brutes have consciousness and no souls, then

> it is clear that, in them, consciousness is a direct function of material changes; while, if they possess immaterial subjects of consciousness, or souls, then, as consciousness is brought into existence only as the consequence of molecular motion of the brain, it follows that it is an indirect product of material changes. The soul stands related to the body as the bell of a clock to the works, and consciousness answers to the sound which the bell gives out when it is struck.[24]

Often overlooked is that Huxley's essay was intended as a limited defense of but also a challenge to Descartes' theory regarding the status of animals as unconscious automata. Huxley, fully committed to the Darwinian theory of continuity across species, rejects the Cartesian separation of human mental life from the balance of the animal economy. He acknowledges that we can be sure of no consciousness other than our own, but he notes that it is in the conduct and manner of others that we plausibly impute consciousness to them. So, too, do animals display the behavior strongly suggestive of consciousness and, in any case, says Huxley, given the suffering they would endure (and have endured) on the assumption that they are but unconscious machines, it is better to err on the generous side.

But what is to be made of the claim that "consciousness answers to the sound which the bell gives out when it is struck"? The relationship between mind (soul) and body, according to Huxley, is akin to that obtaining between a clock's bell and those mechanical works that activate the bell. Thus is sound produced. Now, however—and unlike any mere clock—there is something that *answers* to the sound. This is consciousness and, though it be a mere byproduct—to use Huxley's word, a mere *epiphenomenon*—it is unlike both the bell and the works that activate it. Perhaps it is powerless, lacking all causal power to do anything other than "answer" or witness, but it remains an ontological dangler in that case and therefore still a problem. The central problem, of course, is what Joseph Levine calls the "explanatory gap." To wit:

> Antimaterialists assert that . . . mental phenomena are different in kind from physical phenomena. Among the weapons in the arsenal of antimaterialists, one of the most potent has been the conceivability argument. When I conceive of the mental, it seems utterly unlike the physical. . . . Thus what the conceivability argument demonstrates is the existence of an explanatory gap between the mental and the physical.[25]

Levine accepts the physicalistic alternative to antimaterialism but recognizes the significance of the "explanatory gap" and the need to address it straight away. However, though Descartes did, indeed, note how one might conceive of any number of conscious experiences without requiring the concurrent conception of a body, the gap does not depend primarily on conceivability. It is more centrally associated with the ineradicable difference between any and every physical feature and any and every feature of consciousness as such. Again, the gap separates not two realms of the imaginable but realms of two qualitatively distinct objects of knowledge. At still another but comparably fundamental level, the gap separates the epistemic corrigibility of those witnessing any aspect of physical reality with the epistemic incorrigibility of those "witnessing" (possessing) their own conscious states. There is even more to the "gap" than this, including just what would count as an "explanation" in considering matters of this sort. These several aspects of the "gap" are taken up later.

Emergentism, supervenience, and epiphenomenalism, which, collectively, deny the immateriality of consciousness, offer no argument or evidence that seriously challenges dualistic alternatives. It cannot even be said that they are working hypotheses, because a working hypothesis is one that will rise or fall on the basis of relevant evidence, and there is no "evidence" as such that could tell for or against "hypotheses" of this sort. They are not scientific hypotheses, though they are phrased as if they were. No one disputes the ultimate materiality of oak tables or of water. Consciousness just lacks the "feel" of materiality, so examples drawn from the realms of dead matter are unconvincing.

But perhaps consciousness, this ubiquitous feature of the waking state, should just be taken as a "given." This is a position endorsed by Alexander Bain and recently defended by David Chalmers. We simply postulate consciousness or conscious experience as a feature of reality, akin to mass or charge, and accept that it is not reducible to some more simple or basic congeries of parts.[26] Chalmers has attempted to preserve the scientific options for the study and explanation of conscious experience while acknowledging its immunity to reductionistic strategies. He writes:

> I suggest that a theory of consciousness should take experience as fundamental. We know that a theory of consciousness requires the addition of *something* fundamental to our ontology, as everything in physical theory is compatible with the absence of consciousness. We might add some entirely new nonphysical feature, from which experience can be derived, but it is hard to see what such a feature would be like. More likely, we will take experience itself as a fundamental feature of the world, alongside mass, charge, and space-time. If we take experience as fundamental, then we can go about the business of constructing a theory of experience. . . . This position qualifies as a variety of dualism, as it postulates basic properties over and above the properties invoked by physics. But it is an innocent version of dualism, entirely compatible with the scientific view of the world.[27]

This view might be taken as a sign of resignation, as a courageous admission of limitations, as an echo of William James's mysterious ontology that takes *pure experience* to be the stuff of reality. However it is taken, Chalmers' own attempt to physicalize or naturalize the theory by attaching it to the metric of information suffers from the same defects he otherwise correctly identifies in alternative accounts. Once the experience is translated into something that is not experience, the scientific advantage gained (if any) is at the price of the phenomenon itself. There is much to be said for accepting that experience as such is just a piece of reality, something that is in the world. However, it remains unclear as to what would follow about the world just in case consciousness were not part of it. Of course, what would follow, just in case mass were not part of the world, is—there would be no world! Moreover, consciousness is just not like the rest of it, however "it" is to be understood beyond the range of the mental. It is rarely a successful strategy, when attempting to account for the complexities of the world, simply to stipulate into existence whatever is needed to fill an explanatory or ontological gap.

This is at the heart of Daniel Dennett's criticism of positions such as that taken by Chalmers. Dennett regards such theories as cut from the same cloth that supported versions of vitalism for centuries.[28] The vitalists, argues Dennett, could find no more convincing explanation for the processes of body than one based on a "vital" principle not reducible to mere materiality. What better way to explain growth, life, death, etc., than in terms of a vital principle, an *élan vital* or *vis vitae* itself not physical and not reducible to matter? Rejecting this innocently sober approach, Dennett reminds us that "life" is a collection of functions achieved by specific organs and processes, each fit for scientific study and explanation. There is no reason in principle to consider consciousness in any but comparably functional terms.[29]

Dennett's defense of functionalism is based on the rejection of an alternative solution to the mind/body problem, the "identity theory."[30] Instead of accepting the reality of the mental and then attempting to establish the means by which it may be causally brought about by or emerge from the physical, identity theorists argue that mental terms, if they are referential at all, refer to physical events in the brain. To report a sensation, then, is not to refer to something ineffable and mental but to report what is in fact a state or process in the brain, whether the percipient knows it or not. To report "lightning" is to report electrical discharges, though for centuries reporters knew of no such things as electrical discharges. Nonetheless, if the utterance "Lightning!" refers, it refers to electrical discharges.

Identity theories come in different forms, the most common in philosophy of mind being so-called *type-type*, *type-token*, and *token-token*. Where the purported identity relation is between descriptions or designations of the same individual item, the relation is said to be that of *token* identity. Where, instead, the purported identity relation is between the same kinds or types of things, the relation is that of *type* identity. Regarding the things of the world, one kind or type of thing is fruit. A token (instance, example) of this type is an apple. Regarding the physical events occurring within the body, one type or kind of such events are events in the brain. A "token" of this might be electrical discharges in the optic nerve. With respect to so-called mental events, one such type is what we call sensations. One instance or token of this type might be "red." With such a scheme, a number of mind/brain-identity possibilities obtain. A theory may contend that each type of "mental" event is just a type of brain activity. Thus, thoughts as such are just neural discharges as such. Or the theory might require that a given type of mental event be identical to a specific neurological token: thoughts are *specific cortical discharge patterns*. Or a "token" mental event (thinking of the smallest prime number) is identical to a "token" brain state: discharge patterns in these specific cortical neurons. The overall scheme eliminates the reality of the "mental" and leaves as a residue only an anachronistic mode of discourse fashioned long before science had discovered the operating principles of the nervous system. We still say "Lightning!" now knowing, of course, that it is electrical discharges we're reporting.

Perhaps the more defensible version of the identity theory is the token-token variety, for this is the one that would permit actual and specific measurements and observations. After all, to refer to "brain processes" is so indefinite and airy a reference as to be scientifically useless. What is counterintuitive in the token-token scheme, however, is the specific hardware requirement: If any sort of "mental" event (e.g., thoughts of the smallest prime) must be identical to a specific neural discharge pattern, then no two (different) brain states would permit the given thought to persist. The same scheme seems to legislate against identical sensory or perceptual events in creatures with even slightly different nervous systems, and this runs counter to behavior routinely displayed by these same creatures under comparable circumstances. For a number of philosophers and philosophically interested scientists, it makes more sense to think of the "mental" outcomes as the output of systems that might differ significantly as "hardware" but that are regulated by comparable "software" programs. Put another way, one can do sums by counting one's fingers and toes, by writing with pen and paper, or by operating a calculator. In all such achievements one is *functioning* as an adding device, and this remains so despite the very different operations and machinery employed.

Consider a far more complex set of functions as given by the following equation:

$$Pr = PtG^2 \Theta H \pi^3 K^2 L / 1024 (\ln 2) \lambda^2 x Z / R^2$$

This bit of mathematical doggerel is the radar equation: The power of a signal returned (Pr) by a target is a function of a number of variables, such as antenna gain (G), the angular beam width of the radar signal (θ), and the wavelength of transmitted energy (λ), etc. Any system that satisfies the terms of this equation just is a radar system, and it would not matter in the least whether the components were made of metal, liquid, or gas. What makes the system a "radar" are the *functions* realized through its operation. So, too, the animal, as a collection of integrated systems, will function in such a way as to metabolize food, reproduce, engage in energy exchanges with the environment, construct dwellings, find protection from the elements, etc. Just what machinery is involved in performing these functions may be of interest in its own right, but the proper characterization of the system is in *functionalist* terms.[31] Presumably, in time such functions will come to be reduced to precise formulas. What the Krebs cycle is for carbohydrate metabolism, some other set of functional equations will be for consciousness, or so the functionalists contend.

In *Brainstorms*, Daniel Dennett has offered an account of consciousness based on just this sort of operation.[32] Noting that a conscious inner (Cartesian) self creates the problem of an infinite regress (for that self would then need its own inner self, *ad infinitum*), Dennett proposes an entire cadre of thoughtless little *homunculi*, each carrying out a small portion of the overall project of generating the functions we regard as mental. No one of them alone has any mental life, but all in some sort of synchrony achieve the end. By analogy, think of how many separate steps are needed for DNA sequences to produce a complex protein or enzyme.

This and similar analogies are more reassuring than convincing, however. The stuff of DNA is not unlike the stuff of enzymes. The stuff of neural hardware seems utterly unlike the stuff of consciousness and is not likely to become more like it by adding further hardware. At its foundation, Dennett's thesis is a composite of emergentist and modularity theories, neither of them alone or in combination yielding measurable progress toward a fuller understanding. The actual person who tastes the lemonade and listens to the rolling surf has no access to the "stupid homunculi" but must have some means by which to coordinate and integrate their marginal little achievements into the full-blown reality of mental life, as Aristotle might have reminded us. As it is *that* achievement that constitutes the problem of consciousness, metaphorical metaphysics at the level of the little people is not helpful.

In *Consciousness Explained*,[33] the same *lacunae* separate otherwise informative pages devoted to philosophy of mind. Here Dennett accepts first-person reports of conscious experience but takes them to be data rather than evidence; that is, the reports are not the last word on the reality of an inner life but the outputs of a complex system whose operations require further analysis. Dennett's own characteristically metaphorical analysis is that the physical processes in the brain are akin to the multiple drafting functions that lead to a final and integrated work, here the final version arising from something of a Darwinian struggle. Merging Dennett's notions produces the strange spectacle of a disordered army of mentally deficient *homunculi* fashioning little bits of text, pulled together (somehow) to reach the level of drafts, these then (somehow) "selected" on the basis of their survival value. This is offered as a scientific improvement on "Cartesianism." Needed here is some metric by which to gauge progress.

Constant in Dennett's work is this emphasis on *function* and functionalism, but "function" is one of those terms of art that, taken at face value, often prove to be *ipsedixits*. To say that a system "functions" in such a way as to "realize" a given "software program" is either to offer one of an indefinitely large number of metaphors useful in conveying how one might wish to have something understood or to advance an explanation for behavioral or cognitive outcomes of a certain kind. An animal deprived of food begins to eat once food becomes available. It is not a "brain" realizing a program but an animal engaged in behavior more readily predictable from the conditions of external privation than from any fact about the "function" of the animal's brain. Nor is it informing to refer to "brain" function, for "brain" is, in fact, a constellation of distinguishable nuclei, tracts, chemical pathways, synaptic interconnections, ever "on," ever changing. To suggest that a given "type" of mental event is identical to a given "type" of brain event is to match a fairly stable event (how tall I see my friend as being) with something utterly unstable (the sensory signals produced as my friend moves in space to greater or nearer distances, up and down various inclines, through mist and rain, even as I, too, move). The constancy phenomena (size, shape, color) are but a fraction of the very large number of phenomenologically stable experiences obtaining under conditions of constant neurophysiological and environmental flux. If the participating homunculi are ignorant to begin with, reality must drive them to drink!

A variant of Dennett's approach is that of *Conceptual Role Semantics* (CRS), as advanced by Ned Block and others.[34] On this account, the external world of physical objects, to be objects of consciousness, must gain entry to the system by way of representations: neural codes, synaptic networks, etc. At the same time, for these representations to have any part in mental life, they must carry some meaning. According to CRS, a representation's meaning is just the role it plays in one's overall cognitive life. One who is thinking or perceiving or believing is engaged in an activity with content. The content is just that ensemble of representations. Their respective "meanings" are just their respective functions in grounding the activity itself. As the meaning of a word is established by its use among conversationalists, so the meaning of a representation is how it is used in a brain, a network, a computer.

Block, like Dennett, is a functionalist and has attempted to demystify consciousness by translating it into causal roles at the level of the system's performance. John Searle's still suggestive "Chinese Room" criticism is one that Block and others regard as unsuccessful as a counter to functionalism.[35] Searle treats the computer model of mental life as failing to deal realistically with the central issue of *meaning*. He exemplifies this with the now famous thought experiment that finds a person totally ignorant of the Chinese language given the task of arranging cards according to a set of directions. The cards actually contain Chinese characters (of which the card sorter is ignorant), and the sorting directions say no more than to place a card bearing such-and-such a pattern here or there. The task completed, the sorter can say nothing as to what the series of cards "means," or even if there is any meaning at all. However, native speakers of Chinese, examining the set, immediately comprehend the meaningful statements it conveys. Against functionalism, Searle argues that the brain-as-computer is but an elaborate card-sorting device with "meaning" left unaddressed.

Can the functionalist answer this criticism effectively? Here is how Ned Block answers it:

> If we can program a computer to be intelligent, it won't be the central processing unit (CPU) all by itself that is intelligent or that understands the symbols, but rather all the complex relations between the CPU and other subsystems of the mind, e.g. for perception, reasoning and decision making. So the whole system understands Chinese even if the person who is simulating the CPU does not.[36]

This cannot be an adequate rejoinder unless "understands" is defined in a most unusual way. What renders the sequence of cards meaningful to native Chinese speakers is not "complex relations between the CPU and other subsystems of the mind," for, if this were so, then the card sorter's failure of comprehension would be explicable in terms of the "complex relations between the CPU and other subsystems of the mind." But this is not the source of ignorance. Presumably, a healthy card sorter hosts healthy relations between and among the structures of the nervous system. Lacking is *knowledge of Chinese*, full stop, and the explanation of this, as well as its repair, requires no recourse to CPUs or other subsystems. To be avoided, as we strive toward self-understanding, is that form of superstitious thinking that reifies what at base is an arguable metaphor.

To be sure, there is not much harm done and perhaps some gain in insight or clarity in considering consciousness to be a "function" realized by the brain as a whole, as a complex system. Surely consciousness understood as requiring a functioning brain as a necessary condition is not only a reasonable hypothesis but one for which there is a broad and deep evidentiary support. This, however, does not solve the "problem" of consciousness if what is problematic pertains to the unique ownership and the phenomenological features of consciousness as such. To know that these features require a functioning brain as a necessary condition is not unlike the requirement of ink for the process of printing. That ink is a necessary condition leaves utterly unaddressed what the contents of various compositions will be.

To consider notions of ownership and phenomenology is to raise the now familiar question in philosophy, the question of "what is it like . . . ?" *What is it like to be a bat?* asks Thomas Nagel in a widely cited article intended to underscore the problem of consciousness.[37] For Nagel, the radical differences between human and bat physiologies are such that we could not know what it is like to be a bat, nor, presumably, could bats know what it is like to be us. Moreover, owing to the very phenomenology of experience itself, standard reductionist strategies for settling the mind/body problem simply miss the point at issue. Nagel frames the issue with clarity:

> Without consciousness the mind-body problem would be much less interesting. With consciousness it seems hopeless. The most important and characteristic feature of conscious mental phenomena is very poorly understood. Most reductionist theories do not even try to explain it. And careful examination will show that no currently available concept of reduction is applicable to it. Perhaps a new theoretical form can be devised for the purpose, but such a solution, if it exists, lies in the distant intellectual future.[38]

What Nagel recognizes is that consciousness introduces a new ingredient in the perceptual transactions between organisms and environments. The ingredient is the actual state of experience itself, what may be called a "mental" state presumably widespread in the animal kingdom. The condition that must be satisfied, on Nagel's account, for an entity to possess such a mental state, is one that establishes that "there is something that it is to *be* that organism—something it is like *for* the organism."[39] And, as the standard reductionistic accounts of the mental are essentially indifferent to the subjectivity of such experience, the accounts are fatally incomplete.

Any number of rejoinders to this position have been advanced, though the main point of Nagel's argument retains its power. It is sufficient here to consider briefly David Armstrong's analysis of consciousness, which leads him to conclude that the seemingly special nature of "introspective consciousness" (which, presumably, grounds whatever it is to be *like* what one is) is no embarrassment to the physicalist. This special mode of consciousness, Armstrong argues, is what integrates the various perceptual inputs into some sort of unity that is then taken to be the achievement of some sort of "self." But "a Physicalist, in particular, will take the states and activities introspected to be all physical states and activities of a continuing physical object: a brain."[40]

What the rejoinder misses (as do all kindred attempts to base on physical or mechanistic grounds the phenomenology of "what it is like to be something") is precisely that feature of introspection that carries with it "what it is like to be. . . . " This much of Nagel's analysis seems unexceptionable and still timely. But he proceeds in his analysis toward conclusions that are rather less convincing. The bat is useful to Nagel owing to its navigation by sonar with an auditory system whose operating range is far beyond human powers. The conclusion reached is that we (therefore) cannot know what it is like to be a bat and (therefore) would not be able to specify the required physical correlates or causes generative of the relevant mental states. Implicit in this thesis is that knowledge of what it is like to be in a given mental state thereby permits at least in principle the identification of those physiological processes on which such states somehow depend. There is convergence here of Nagel's and Armstrong's positions. Also implicit in the thesis is the assumption that one who is in such a mental state "knows what it is like." However, neither of these assumptions is warranted; indeed, neither lends itself to credible modes of verification. If one inhales deeply the emanations of a nearby bouquet, there will be the experience of something fragrant. There will not, however, be the experience of being in a mental state, for what is experienced is what is given off by the flowers. If, in the circumstance, one knew that one was in some sort of "experiential" mental state, there would be no basis on which to assume that a complete physicalist account of that state would be an account of the fragrances. Put another way, knowledge of one's mental state (if such is possible) would be devoid of the very *qualia* that the physicalistic account is supposed to explain.

More confusing is the notion of knowing "what it is like to be" in one or another mental state. One could report, in an unusual way of speaking, something like, "I know that I am in the mental state of smelling the lilies." It is unclear, however, what then could be made of the query, "How do you know this?" There may be some philosophical distinction to be made between "experience" and "knowing what it is like to have experience," but there is surely no phenomenological distinction. It is precisely because first-person reports of experience are, as it were, incorrigible that they are not "knowledge" claims in the ordinary sense. One does not know "what it is like to smell lilies" over and against smelling lilies. Moreover, it is far from clear that radical differences in physiological processes and anatomy inevitably yield radical differences in "what it is like to be" one or another creature. For this to be clear would require the very reductionistic successes that are at issue.

From species to species, a flower may be perceived as yellow or as orange or as an ultraviolet mass—in all these cases having something akin to the quality dubbed "color" by most of us. Similarly, Bob, the dichromat, has defective color vision and sees the red rose as gray, though his brain state is very similar to the one taking place in Helen, the trichromat, who sees the rose as bright red. Note, then, that similar brain states might yield different "mental" outcomes, whereas very different nervous systems might function in such a way as to produce comparable outcomes. Clearly, it is not to be expected that a given type of experience matches up with a specific type of nervous activity (or nervous system). That is to say, what is called a type-type identity must surely be the exception. There seem to be few brain states so stable and replicable as to be able to account for the stability of any number of mental states.[41]

As a necessary aside at this point, some reservation is called for regarding this now standard way of expressing possible grounds for mind-brain identities. The principal terms refer to "types" and "tokens," but it is clear that neither term has sufficient stability in application to leave readers unperplexed or, in the end, to do the work required in addressing serious metaphysical issues. Consider the large and varied *types* of transactions by which some part of the wealth of one person (family, tribe, nation) is conveyed to another. We might bring all of them under the "type" *Last Will and Testament* as distinguished from the different "type" *Transfer of Funds*. One "token" of the former would be Janet's will; of the latter, Janet's personal check. But both her will and her personal check are also "types" of instruments of which there may be very different "tokens"; for example, a will dictated, signed, and notarized, versus one uttered on one's deathbed and witnessed, versus a recorded statement finally accepted by a judge of the probate court. A "personal check" is really any written document a bank will accept, even one penciled in on a restaurant napkin. As I say, "types" and "tokens" are far too ad hoc for serious metaphysical purposes.

All this duly noted, some contend that a given "type" of mental state in Martin (e.g., Martin in an "experiential" state) just is (i.e., is identical to) a "type" of activity (e.g., increased electrical activity) in Martin's brain. The match here is between a mental state of a certain "type" and a brain state of a certain "type." (In actuality, we rarely learn just what "type" the philosopher has in mind, for both the "mental" and the "brain" states fall under the airy category of "states" or "processes," encouraging readers to exercise their own imaginative powers.)

Then there are proposed *type-token* identities, where a given type of brain state (increased electrical activity) is presumably identical to *this* (token) mental state for Sam but *that* (token) mental state for Pam. The greatest specificity is that which collapses a mental *token* and a neurophysiological *token* into an identity relation. Whenever fiber J of the VIIIth cranial nerve, activated by hair cells along the basilar membrane, generates rate R in impulses per second, Pam and Sam report hearing the tone middle C. Thus (though this surely does not follow), the auditory experience of middle C just *is* R impulses per second in fiber J of the auditory nerve.

Consider how "type" and "token" figure in the larger debate. It is common to illustrate the difference with words or numbers. The telephone number for obtaining directory information in New York City is (212) 555-1212. How many numbers are there in this set? If each entry is considered as a distinct "token" of what it is to be a number, then there are ten numbers. But note that there are four 2's, three 1's, and three 5's. There are, then, only three *types* of numbers. We might also say there is actually only one token of the type *even numbers*. Or, again, we might correctly claim that there are six "tokens" of the odd numbers and four of the even numbers. See how elastic is this division of items into types and tokens!

What, then, of "type-token" identity? This attempt to save identity theories is otherwise overburdened by conceptual problems. What is asserted is that, for the specific person or animal, a given type of mental state (smelling) can be achieved by way of quite different (tokens of) brain processes, though, whenever this type of (olfactory) experience occurs, it is identical to specific token events in the brain. Of course, the same specific token brain processes in two different persons might yield quite different "types" of olfactory experience. However, admitting this much finally leads to a rejection of type-type and token-token identities while leaving type-token identities so ad hoc as to be jejune. The entire framework seems to be a form of scientific special pleading, the unwanted mutant of a failed breeding program. What can be made of a theory that admits of various (unspecified) "mental" instances arising from something standing as a loosely defined "type" of brain "activity"?

Let me return to the fairly basic and established relationship previously cited. As the intensity of light reaching the retina of primates is increased, there is a measurable increase in the rate at which fibers in the optic nerve emit electrical pulses. Concurrent with these changes are alterations in behavioral or verbal indices of the perceived brightness of the light. Based on such findings, two versions of the identity thesis would seem to find some support. Taking the neural changes as falling under the "type" *magnitude of response*, and taking the perceived brightness changes as also instances of *magnitude of sensation*, we might accept something of a *type-type identity*. Then, too, it would surely be permissible to say that for a given *type* of neural activity (increased rates of "firing") there is the associated *token* of the mental, the experience of brightness of a magnitude of so many units. But here we have no more than the well-established psychophysical law (Fechner's law) describing the relationship between stimulus magnitude and sensation magnitude. There is no "identity," as such, merely a reliable relationship. To establish any relationship, it is necessary to have terms on both sides of an equation, but in the absence of the *experience* of brightness, there is nothing to which stimulus intensity can be related.

The persistent problem of *qualia* enters here with a vengeance. The experiences of brightness, loudness, color, taste, etc. are in no way like discharges in sensory nerves or changes in the electrochemical processes of the brain. To this extent, *type-type* matches are out of the question. It would strain credulity to suppose that for every conceivable type of mental event (e.g., *having the property of believing* that tax cuts are good for the economy) there is a matching type of physical state (e.g., the nervous system *having the property of* "X-ing"). Besides all this, there is something contrived about expressing things in such a manner. We know what we mean when we declare that Elizabeth believes running is good for the heart. We would be inclined to think it a trick question if asked: "Does Elizabeth *have the property of believing* that running is good for the heart?" One might say without ambiguity that the fullest description of Elizabeth would include her beliefs and, in this somewhat pale sense, we might agree that her beliefs are "properties" or attributes. However, the basis on which we attribute beliefs to Elizabeth is not the basis on which she holds them. That is, Elizabeth does not resort to a table of descriptors in order to discover whether it would be correct to declare herself to have the property of believing that running is good for the heart. She just believes it. She and we, however, would certainly require much by way of evidentiary resources to uncover some "type" of brain state—*any* type of brain state—putatively matching Elizabeth's beliefs. Three nonmatching epistemic routes are required here: (1) the short one, used by Elizabeth, which just is her belief; (2) the inferential one, used by us, in concluding with less than certainty that she does, indeed, believe running is good for the heart; and (3) the winding and perhaps finally aimless one we and she would use to locate something in her nervous system that offered evidence or mere correlates of *having the property of believing*. Compared with this mode of analysis, *token identity* becomes nearly credible, even if no more successful at the level of explanation.

So far, it seems as if there really is this problem of consciousness and that conventional approaches to it have turned up little beyond what was on offer as Descartes submitted to the criticisms of his contemporaries. Their criticisms were at least as discerning as those produced by current anti-Cartesians, and as the criticisms were answered at length by Descartes, perhaps "Cartesianism" is worth another look, though surely not for the sake of reviving it, chapter and verse, or for accepting the implications Descartes believed to be warranted. "Cartesianism" is worth another look chiefly to see just what is taken to be a good reason for rejecting a theory and what is taken to be the gains associated with newer formulations.

3

"CARTESIANISM" REVISITED

Both of these statements can't be true:

(1) "*Why Isn't the Mind-Body Problem Medieval*? ONE ANSWER: Because medieval philosophy is just the continuation of ancient philosophy by other means . . . and . . . the mind-body problem isn't ancient."[1]

(2) "In his discussion of sense-perception, Augustine shares the preoccupations of philosophers in the Greco-Roman tradition, and the problems which he identifies remain in many cases those of the modern philosophy of mind."[2]

As noted in the first chapter and notwithstanding to the contrary the contributions of Plato and Aristotle, scholars are still not in agreement on the question of just when and for whom mind-body relations were recognized as metaphysically problematic in the way they are in modern philosophy. That there was widespread recognition of the special nature of the "mental" over the various seasons of philosophical speculation is beyond doubt. Centuries before Descartes and "Cartesianism," philosophers had devoted whole treatises to the mind-body distinction and division, often in a fashion close to Descartes' own. After Plato and Aristotle, the Christian world of thought engages the problems in the writings of Paul and especially of Augustine. The issue becomes ever more insistent in Scholastic and Renaissance philosophy when Aristotle's authority seems to stand behind the mortality of the soul. All in all, one or another version of the mind-body problem is addressed throughout the history of philosophy.[3] Perhaps what is salient about Descartes' place within this veritable tradition is that it falls within a self-consciously scientific age now largely won over to materialistic conceptions of reality. In light of the attention given to the mind-body problem by a small army of important commentators that would include Plato, Aristotle, Galen, Augustine, Aquinas, all the Renaissance Platonists and Aristotelians, and then Francis Bacon, it seems as unfair to pick on Descartes as it is unfair to intellectual history to credit him with originality on this particular point. On the more specific matter of consciousness, this too, as a term and as a phenomenon, manifests itself variously in different epochs of philosophical discourse. Here again it is especially irksome when Descartes is so routinely cited either as the first to see the problem in its philosophical depth or to frame a solution that would (by certain lights) be regarded as simplistic or even incoherent.

Having referred to "Cartesianism" several times already, noting the broad range of ascriptions adopted by its critics, it is time to introduce some precision. I should think it especially useful to recall how "Cartesianism" was understood later in the seventeenth and early eighteenth centuries, and chiefly by the more humanistic scholars of the period. For them the problem posed by Cartesianism was not a two-substance ontology but an utterly *mechanistic* account of reality. Descartes' philosophy surely does include a two-substance ontology ("mental" and "physical"), but it is defended by what turns out to be a fairly common-sense account of properties that makes sense to most people even today. Thoughts have nothing in common with things. The latter are massive and extended, may be hard or soft, have a shape and occupy a specific place. Thoughts are different, and substantially so. This is all too obvious to call for centuries of argument. More vexing to such important commentators as the Cambridge Platonists was that feature of Descartes' system that stood in substantial agreement with the Newtonian worldview, Locke's philosophy, and Galileo's "new science." They all reduced reality to *mechanism*. Henry More, a leading figure within Cambridge Platonism, protested as early as 1668 that "it is as confessed a Principle with him, that matter alone with such a degree of motion as is supposed now in the Universe will produce all the phenomena of the World, Sun, Moon . . . Plants, Animals, and the Bodies of Men. . . . "[4]

But Descartes' common sense takes him further to the point at which he must also acknowledge an interdependence between these distinct ontological realms, for it is clear that some thoughts lead to bodily movements and some changes in the body clearly have effects on mental life. The result of this at the metaphysical level is, to use Jaegwon Kim's useful expression, "that under Cartesian dualism there can be *no complete physical theory of physical phenomena*."[5] The problem, then, as Kim says, is that "*Cartesian interactionism violates the causal closure of the physical domain*."[6]

So much of the artistic and religious thought of the period was inspired by a neo-Platonist perspective that this Age of Newton created an intellectual battleground on which the major contestants were mechanists and Platonists. If the former may be said to have a patron saint, it was surely Hobbes, whose mechanistic theory of mind was inspired by the achievements of Galileo and fully anticipated Locke in the matter of the source of mental contents. Early in *Leviathan* Hobbes leaves no doubt but that "there is no conception in a man's mind, which hath not at first, totally, or by parts, been begotten upon the organs of sense."[7]

Samuel Mintz begins his detailed examination of the reactions to Hobbes within the mechanistic philosophies of the seventeenth century with this summary indictment:

> Hobbes was the *bête noire* of his age. The principal objection to him, the one to which all other criticisms of him can ultimately be reduced, was that he was an atheist. He was the "Monster of Malmesbury," the arch-atheist, the apostle of infidelity, the "bug-bear of the nation." His doctrines were cited by Parliament as a probable cause of the Great Fire of 1666. His books were banned and publicly burnt, and the ideas which Hobbes expressed in them in his lucid and potent style were the object of more or less continuous hostile criticism from 1650 to 1700.[8]

Even within the pantheon of seventeenth-century science there were principled objections to the radically mechanistic and atheistic character of Hobbes's philosophy. Robert Boyle, already engaged in an irksome exchange with Hobbes, added his illustrious name to the list of good Christians who found Hobbes's science as weak as his philosophy.[9]

For Hobbes surely, physics is complete and, if physics is complete, all causal laws are at bottom laws about purely physical interactions. Once these are included, the causal account is closed. But if among links in the causal chain or ancestors in the causal pedigree there should be a thought or two, a great and grave problem arises; viz., the causal closure of the physical domain is violated. What is not considered, before seeking to prosecute violators, is the possible wrong-headedness of ever thinking in the first place that everything that matters is finally explicable in physicalistic terms. The idea itself is an echo of *Enlightenment* hubris as expressed by Laplace in his *Philosophical Essay on Probabilities:*

> We may regard the present state of the universe as the effect of its past and the cause of its future. An intellect which at a certain moment would know all forces that set nature in motion, and all positions of all items of which nature is composed, if this intellect were also vast enough to submit these data to analysis, it would embrace in a single formula the movements

> of the greatest bodies of the universe and those of the tiniest atom; for such an intellect nothing would be uncertain and the future just like the past would be present before its eyes.[10]

Left out of this confident account is the intellect itself, which is either nothing but a moment in the physical chain of positions and motions (in which case it cannot go beyond its own final link) or it is outside this very chain. Laplace's "demon," an intellect rich in data and computational agility, might well reduce to a single formula all of cosmic dynamics (with Heisenbergian uncertainties dealt with somehow). If the demon also possessed the power of self-reflection, however, it would come face to face with "Cartesianism," would begin to wonder just how atomic forces and positions generated the visual *qualia* "before its eyes," and would then reconsider whether there is likely to be the "causal closure" of physicalism. As Shaftesbury put it, good Epicurus would finally have to tell us "how atoms came to be so wise."[11]

In the first chapter, I offered accounts of the approach to such problems taken by Aristotle and Plato, followed by contemporary attempts at defining and solving them. Before returning to Descartes in this chapter, two related matters require more than a brief discussion. The first has to do with philosophical methodology, specifically, the fate of the mind-body problem and the problem of consciousness once a given philosophical *method* becomes more or less official. This is abundantly clear in Descartes' own and explicit commitment as developed and defended in his *Discourse on Method.* It is just as clear in contemporary philosophy, which confers methodological authority on modes of conceptual and linguistic analysis. To make the point again, the matter of method is a necessary preliminary issue, for Descartes' philosophy of mind is beholden to his methodology. If the philosophy's putative defects are, in the end, the inevitable result of methodological disagreements with his critics, then the substantive claims are put on "hold" while critical appraisals of competing methods are carried out. As the confrontation of

"Platonism" and "mechanism" unfolded in the seventeenth century, the terms of the contest were still too new for either side to have a retrospective vantage point, just as they lack one in assessing our own conceptual loyalties. As R. L. Brett noted:

> The modern mind feels at home with scientific categories and it is difficult for us to imagine how a seventeenth-century mind that had not embraced the new science would organize and carry out its thinking. . . . The rising tide of scientific thought swept away symbol and analogy. The scientific habit of thinking in terms of cause and effect became stronger than the habit of perceiving symbolic relationships.[12]

The Platonic mode of cognizing physical reality as symbolic and suggestive of higher (transcendent) truths, of realities that are ultimate, was replaced by modes of comprehension that were stridently *mundane* and seemingly vindicated by practical success with problems the world regarded as serious. Descartes within this context would be elevated to the highest ranks of the priests of science and certainly not as a defender of an older metaphysics.

The second preliminary matter arises from intellectual history and, alas, does show how Descartes (as with all well-educated persons) would be tied to an older metaphysics. For all his originality in mathematics, he was after all the product of a philosophical education and a philosophical culture so utterly indebted to medieval rationalism ("Scholasticism") that "Cartesianism" itself is likely to be misidentified outside that context. This education and culture are not widely understood in our own time. Accordingly, I will summarize several of the major developments taking place from the later patristic period to the educational initiatives of the later medieval period. Even this condensed account should reassure readers that philosophically engaged scholars in any age will recognize as a fundamental metaphysical ("Platonic") conundrum the juxtaposition of an utterly physical external world and the seeming nonphysicality of the ideas we form about that world.

What is new in "Cartesianism"? It is time to clear the boards of misleading characterizations, not to mention the odd notion that, with Descartes, something unheralded in the history of philosophy was brought into being. Considering again the ancient Greek thinkers, neither Plato's uncompromising mind-body dualism nor Aristotle's conservative naturalism would enjoy dominance during the productive periods of Hellenistic philosophy. As Heinrich von Staden has noted, "despite their radically divergent versions of the world and of human beings, Epicurus, many Stoics, and the more significant early Hellenistic physicians share a constellation of convictions. . . . That all *psyche* is *soma* . . . that only what is spatially extended, three-dimensional, and capable of acting and being acted upon exists."[13]

Here, then, is the earlier version of the oscillations between dualism and monism that would be repeated in various forms up to the present time. In the period between the decline of Hellenism and the rise of Christianity, the most prominent philosophical schools were decidedly materialistic. Within these schools, consciousness as such was either neglected or was rather ambiguously absorbed into a general psychosomatic framework that included as well the faculties of perception, memory, emotion, and desire. It is the rise of Christianity, with its moral theories of sin and redemption, its insistence that the believer consult what is found in his heart, that marks the appearance of "consciousness" as a subject of now surpassing importance. We will find the first full expression of Cartesianism so called, including a version of the *Cogito* argument itself, fully developed in Augustine's treatise on the Holy Trinity (*De Trinitate*): "The knowledge by which we know that we are alive is the inmost knowledge where the Academic cannot even say, 'Perhaps you are asleep and are unaware of it, and you see things in your sleep' . . . So let a thousand kinds of deceptive percepts be presented to him who says, 'I know I am alive,' he will fear none of these, for even he who is deceived is alive."[14]

This is in book 3, chapter 12 of *De Trinitate*, the chapter in which Augustine levels his critique against the then influential Academic (i.e., skeptical) philosophers.[15] Noting that the senses are often instruments of deception and, as such, offer ample evidence in support of skepticism, Augustine presses on to find sources of knowledge immune to such attacks. His version of the *Cogito* is developed in this connection.[16] He chides the Academic philosophy for, having noted that the insane person regards himself as sane, the Academic philosophy then becomes, "still more wretchedly insane by doubting all things." Obviously, Descartes' antiskeptical productions are a later edition of the Augustinian and avail themselves of a comparable *introspective* mode of analysis.

Augustine respects the power of our perceptual resources, respecting these even in light of their limitations, but also respecting yet other modes of knowing not thus limited. His anticipation of the *Cogito* argument finds him claiming, against the Academics, that

> whereas there are two kinds of knowable things,—one, of those things which the mind perceives by the bodily senses; the other, of those which it perceives by itself,—these philosophers have babbled much against the bodily senses, but have never been able to throw doubt upon those most certain perceptions of things true, which the mind knows by itself, such as is that which I have mentioned, I know that I am alive. But far be it from us to doubt the truth of what we have learned by the bodily senses; since by them we have learned to know the heaven and the earth, and those things in them which are known to us, so far as He who created both us and them has willed them to be within our knowledge.[17]

And the same work includes in the summary-title of chapter 11 the daring thesis that "the Likeness of the Divine Word, Such as It Is, is to Be Sought, Not in Our Own Outer and Sensible Word, But in the Inner and Mental One. There is the Greatest Possible Unlikeness Between Our Word and Knowledge and the Divine Word and Knowledge."[18]

Clearly, it is not by way of semantics that one reaches the Divine, less is it through empirical modes of knowing. Rather, it is in the inner, intuitive, mental world that the ultimate truths are to be found, at least in the sublunary portions of the soul's eternal life. The process is one of reflection, not perception; one of abstraction, not semantics. That language is a tool here is true but trivial. Words as such face an insuperable barrier, on the other side of which is the divine word reached through an inner mental process that can catch glimmers and intimations by reflecting on its own nature and purposes.

Leaping across centuries to the period of Scholastic philosophy, we find Thomas Aquinas adopting a similar position, though one more decidedly indebted to Aristotle. Yet no less an authority than Anthony Kenny supports the contention that in these developments there is still something missing that will not become paramount until the time of Descartes: the role of *consciousness*. Here is how Kenny frames it:

> For Aristotelians before Descartes the mind was essentially the faculty, or set of faculties, which set off human beings from other animals. Dumb animals and human beings shared certain abilities. . . . But only human beings could think abstract thoughts. . . . For Descartes, and for many others after him, the boundary between mind and matter was set elsewhere. It was consciousness, not intelligence or rationality, that was the defining criterion of the mental.[19]

But the Aristotle who has the mind "taking cognizance" has already incorporated consciousness into the framework of mental life, a consciousness that is also present in animals (who possess imagination and desire, for example, though not *deliberation—prohairesis*). Nor in Descartes is the "dividing line" consciousness, for when he sets out to establish the basis on which a human being differs from an otherwise convincing mechanical analog, it is not consciousness that stands as the criterion but the capacity for abstract rationality and language, here adopting the very criteria first developed by Aristotle.[20]

It is often necessary to remind ourselves of the main point of the *Cogito* argument: It was not intended as a chapter within a descriptive anthropology, nor was it intended to vindicate ontological claims as to whether or not there was a "Descartes" to be numbered among the things of the world. The *Cogito* for both Descartes and Augustine is a *methodological* device by which to preserve philosophy from nihilistic skepticism.[21] It is not in virtue of *consciousness* that antiskeptical arguments win the day, for skeptics never denied it. It is in virtue of the incorrigibility of introspective accounts that the complete skeptic is reduced to self-contradiction. Again, I do not prejudge the success of the gambit. Rather, I underscore its purpose in order to reinforce the claim that consciousness becomes a "problem" for philosophy only when philosophers abandon one mode of inquiry in favor of another, and do so in defense of a metaphysics that should be subject to critical appraisal rather than adopted as if self-confirming.

Of course, *if* physics is complete, and *if* this requires us to jettison from reality all that is not physical, and *if* consciousness is claimed to be other than an emergent feature of brain activity, etc., then, to be sure, the "problem" of consciousness is solved. But *if* this last step in the series is either wrong or (as I will argue) not fully intelligible, then we return to the wonderfully controversial metaphysical possibility that physics, alas, is *not* complete. Such a possibility becomes at least thinkable when we examine the sense in which any account is taken to be "complete." It can only be complete according to standards of explanation and intelligibility that presuppose a *cognitive* power capable of constructing from an assortment of facts what is finally an *account*, and one that is "complete" according to criteria that are subject to appraisal, challenge, and rejection. Of course, if one insists further that issues of this sort must be settled *empirically*, the question is instantly begged. But what other method is open to us? And it is this question that provides an entry point into what

Plato, Aristotle, Augustine, Aquinas, and Descartes were proposing in different ways, and what the Cambridge Platonists were at pains to direct against the mechanistic reality of Hobbes.

Descartes was an original thinker but also an educated one, whose mature thought was informed by the discipline of a systematic curriculum. It is more difficult to be "original" when one has grasped more completely the achievements of one's forerunners. It is also more difficult when one has been instructed in the more refined instruments of argument. What, then, of young Descartes' schooling? We can begin with the formative years of his education at the new, elite Jesuit school at La Fleche. It was this education that equipped him with the tools of analysis but that also aroused in him deep suspicions about the adequacy of "Scholastic" modes of inquiry and explanation. He entered La Fleche when he was ten and was then channeled for four intense years through one of the most carefully crafted curricula in the history of education, the famous Jesuit *Ratio Studiorum*.[22] It was designed—and it was designed and redesigned with care and with clear-headedness—to develop the powers of the mind and in the process reveal the implications arising from the profound differences between physical causes and moral purposes. Descartes was born in 1596. By 1599, there were nearly 250 of these Jesuit schools, all of them calling for specialists in scripture, Hebrew, Greek, theology, mathematics, philosophy, and moral philosophy. A youngster admitted to any of them was already recognized as possessing special abilities.

In the first-year philosophy curriculum, eight hours each week were devoted to logic and physics, the latter including close examinations of the methodological differences between mathematics and physics. In the second year, the program in physics continued with cosmology, and to this was added the study of psychology, based largely on Aristotle's works. Later would come metaphysics, moral science, and natural science. Nothing in the program was understood to be narrowly sectarian, for a common humanity was presumed to erase all traditional barriers of sect and party. Accordingly, the writer these young Catholic students were required to study and formally imitate under the earliest versions of the *Ratio* was no one other than the pagan lawyer Cicero. (Descartes later earned a degree in civil law from Poitiers, in 1616.) I offer this biographical snippet to highlight the first stages of intellectual development at which Descartes was instructed in the different modes of inquiry, in the desired hope that all would converge on unified truths, and, tied to this, the extent to which mere opinion and probabilities were patent signs of ignorance.

What sort of philosophy of mind might seem plausible in light of this background? Descartes, as noted, is routinely credited (charged!) with presenting consciousness as a "problem." In an attempt to convey just what Descartes' position was on mental life as contrasted with physical objects and properties, I should make quite clear at the outset that I am not defending or seeking to revive that position. Rather, I hope to show that the "Cartesianism" that confers leading-edge modernity on those who criticize and oppose it bears little relationship to what Descartes actually affirmed, and that what he did affirm is not radically different from what is widely endorsed by today's cognitive neuroscientists. The task here is made far easier owing to the fine analysis recently provided by Desmond Clarke, whose book came to my attention after I had reached conclusions close to his own.[23] Chief among these conclusions is, as Clarke says, "the problem with which Descartes struggled in the 1640s is not much closer to resolution today, despite significant advances in our understanding of the properties of matter."[24]

That philosophy would find Descartes' position problematic would not surprise Descartes himself, for, in the very first part of his *Discourse on Method*, he informs readers that the philosophical methods of inquiry are insufficient. To wit:

> Regarding philosophy, I will say only this: seeing that it has been cultivated for many centuries by the most excellent minds, and yet there is still no point in which it is not disputed

> and hence doubtful, I was not so presumptuous as to hope to achieve any more in it than others had done. And, considering how many diverse opinions learned men may maintain on a single question. . . . I held as well-nigh false everything that was merely probable.[25]

Here we have one judged to be the very harbinger of modern philosophy seemingly rejecting philosophy itself. Of course, what he rejected was a method that could rise no higher than speculations and impressions, a method willing to settle for what is merely probable or, put another way, for anything not patently absurd. Needed was a different method and one that was responsive to the highest standards of philosophical credibility as these were systematically developed within the *Ratio Studiorum*. Again, it was not "consciousness" as such that would characterize his foundational criterion but *certainty* and *clarity* as these are faithfully represented in mathematics.

Initially, Descartes regarded mathematics (which, "above all, I delighted in . . . because of the certainty and self-evidence of its reasonings")[26] as pertaining only to the mechanical arts. In the early phase of his philosophical thinking, he "did not yet notice its real use."[27] What was its real use? The real use was as a model and guide, a standard by which to test progress in any field of serious thought, in any field in which problems are of a nature admitting of solution. As a model, mathematics also furnished numerous examples of the importance of stipulations and the ineliminable requirement of starting points, of what the ancient mathematicians called "common notions." The arguments of geometry cannot get off the ground until such notions are in place. To stipulate that a point is the limit of a line is not to encourage some sort of empirical inquiry to determine if this is the case; it is instead the starting point for the systematic development of a deductive science. And though this deductive science is abstract, the ease and the efficacy with which it comes to be mapped on to the physical world reveals an aspect of that world otherwise invisible to the senses. Here, then, is the background of the *Cogito*.

What of the *Cogito* itself? Reading the authoritative literature that has grown up around this argument for nearly four centuries, one might get the impression that Descartes was just a confused plodder unaware of the gross illogicality of his "discovery." What Descartes himself found in the success of the argument was the vindication of a method built on what late in life he identified as the two principles grounding his entire philosophical project. These are discussed in his *Principles of Philosophy*, first published in 1647 and representing his "*Summa Philosophiae*," as he referred to it earlier in a letter to Constentijn Huygens. The two principles flow directly from the *Cogito*, for it is the indubitable fact of thought itself, over and against any and all merely material aspects of an ever potentially illusory world, that must ground knowledge. The principles are these: "There is a God who is the author of everything there is in the world; further, since he is the source of all truth, he certainly did not create in us an understanding of the kind which would be capable of making a mistake in its judgments concerning the things of which it possesses a very clear and very distinct perception."[28]

The theological principle, though central for Descartes, is not relevant to the present discussion, so I turn to the notion of "clear and distinct perception" and the allegedly incorrigible status of knowledge apprehended with the requisite clarity and distinctness. Note that the claim has to do with being mistaken in judging what is otherwise apprehended with clarity and distinctness. It is a claim that lends itself to any number of interpretations, but I presume here to offer the one Descartes would have endorsed. Surely it is possible (for the delirious, the mad, the intoxicated) to form a firm perception of what is in fact utterly chimerical. What would not be possible is for such a person to admit of being mistaken in judgments made *about* these objects of experience. Say, for example, that John tells Mary that the water in the tank in which he has just placed his hand is quite cold. Mary, who knows that John's hand had previously been on a heating pad, informs John that the water is not cold, but only seems so owing to prior exposure to heat. Mary is correct in explaining why John perceives the water to be cold, but John's clear and distinct experience of the coldness renders his judgment incorrigible. He cannot be "wrong" in reporting the contents of his clearest and most distinct perceptions. He may, of course, be entirely unaware of the causes of these contents or their relation to other things of which he has no clear and distinct perception or any knowledge whatever. The criterion, then, is not "consciousness" but the doublet "clear-and-distinct," as in a mathematical demonstration.

We have here Descartes' well-known and routinely chided dependence on the first-person, where such accounts have a special and protected status. On "Cartesian" grounds, nothing in the realm of the *res extensa* can strip such accounts of this status. Nothing about brains, as such, can be used to "correct" what John feels by way of the temperature of the water. The contemporary cognitive neuroscientist or philosophical materialist might wish to challenge all this as mere "folk psychology," but will not have a credible alternative for accounting for the facts of John's phenomenological life. Hand waving and slogans must be unconvincing here. Charles Siewert makes the point directly: "Surely it matters *in what respect*, and *in relation to what*, the superior theory is superior. So we need to know: in what way and with regard to what is commonsense psychology inferior to neuroscience . . . and why does *that* sort of inferiority require us to dump the first, if we cannot subsume it under the second."[29]

For Descartes, it was not a matter of technology that rendered accounts based on materiality here subject to doubts to which the clear and distinct idea is immune. Indeed, those opinions, theories, and speculations, whether in science or philosophy, competing for acceptance must present themselves to the court of judgment, where the evidence that counts is in the form of the clear and the distinct. The doctrine that would have physics "complete" finds its defense in those predictive laws by which we form the most precise understandings of the physical world. It is in the mathematical shape of the developed sciences that we validate our confidence in their claims and methods. In these fundamental respects, the certainty achieved by way of the *Cogito* differs from all merely empirical modes of knowing and is evidence of something about the knower that could not be reduced to the level at which empirical inquiry functions. What is knowable through the operation of our own materiality is ever subject to doubt; what is known with the unmatched clarity and distinctness of conscious experience is not. Thought as such embraces the reality behind the pictures, the plot and point. Were it no more than a copy or echo of impinging stimulation, it could rise no higher than a congeries of impressions. Only an entity that is a *thinking thing*, only a *res cogitans*, can fashion an essentially scientific, systematic account and subject it to the court of thought that must determine whether the claims reach the level of what is clear and distinct. Alas, this is the "Cartesian" *two-substance* ontology!

It is this ontology that has attracted such sustained criticism as to leave the contemporary witness confused as to how it ever was taken seriously in the first place. But today's "contemporary" may prove to be tomorrow's stick figure. The predominant approach today to matters of the sort under consideration is analytical but in a manner different from ancient modes. A fresh look at our own methodological prejudices is useful here. As noted in the first chapter, Plato and Aristotle both possessed authentic credentials as "analytical philosophers," if the term indicates respect for weighing the coherence and meaningfulness of propositions and a deep understanding of how the very phrasing of a problem can create illusory problems and illusory solutions to them. However, the aims of Plato and Aristotle went far beyond "linguistic analysis" and presupposed the possibility of actually solving problems and explaining the nature of things. Recognizing the limits of any and every tool of observation and reason, both were also prepared to accept that some facts were beyond the ambit of human powers but stood as facts nonetheless. That the cosmos as then known offered evidence of design was a fact of daily experience, though neither natural science nor philosophy could render a full and final explanation for it. To put the point simply, the ancient philosophical canon left no room for the comforting belief that what could not be explained philosophically could not be trusted.

Descartes' education at La Fleche was disciplined and thorough, but it did have at least one limitation in presenting the work of the ancient Greek philosophers: the texts were all in Latin! This would work no hardship on students whose lives would be less plagued by philosophical issues than was Descartes'. But for one who would challenge or depend upon core precepts in the writings of Aristotle, there would be room for misapprehension. Perhaps this is nowhere more evident than in Descartes' attempt to work out the sense in which self-movement is to be understood.[30]

Thought of as more faithful to scientific principles than was Aristotle's account, Descartes' position on the self-motion of animals was mechanistic. Indeed, it was Descartes more than most of his contemporaries who was inclined toward the completeness of physics—toward *physicalism*—and it was this very inclination that found him opposing the "Aristotelians," who seemed to require some sort of nonphysical agent as the source of animal movement. Instead of requiring such a "soul," the moving creature on Descartes' account was but a machine with a musculoskeletal organization and fluid dynamics of the right sort. Thus, in his *Discourse* he rejects the "Aristotelians" (here referring to the La Fleche commentators on Aristotle) who would explain movement as soul dependent, insisting instead that the question itself is an empirical one and tied to "the mere disposition of the organs."[31] That this may have challenged the views of "Aristotelians" was all quite beside the point of what Aristotle himself was concerned about. As noted in chapter 1, "motion" in Aristotle is generic for *change* in all its forms, not all of which involve movement as such. As Sarah Byers has revealed in her close study of the issue and in her recognition of Aristotle's theory centering on those motions that ground *metabolism*,

> Descartes interpreted "self-motion" as a reference to the local motions of the constitutive parts of the body. . . . Owing

> to this misinterpretation, Descartes never actually addressed the Aristotelian theory, and so his physics leaves unresolved a problem that Aristotle is able to solve by positing the soul as the life of the body, namely: Why do some things metabolize, but others do not?[32]

This much should make clear that Descartes' confusions as to the traditions he was attempting to overturn arose more from his physicalistic and mechanistic commitments than from anything found in the stock notions of anti-Cartesianism.

In our own time, philosophical investigation has taken that celebrated "linguistic turn," this being part of the mixed bequest of logical positivism and then the more robust bequest of a veritable regiment of "analytical" philosophers. One need only cite in this connection Wittgenstein, Ayer, Russell, and Ryle, followed by a legion of disciples and exegetes. Within this newer tradition, early versions of "verificationism" sought to install empirical methods as ultimate, rejecting as a species of (literal) non-sense whatever fell beyond their reach. As the recognition grew that much of what actually does fall beyond the reach of observation plays some part in the affairs of the real world, the restrictive canon was expanded to include inquiries into cultural and linguistic traditions by which whole worlds are fashioned: worlds of law, aesthetics, morals, politics, religion. To the extent that the contents of such worlds have no direct empirical referents, their sources are to be found in the narratives of a people; storied lives enriched by homegrown realities essential to the preservation of the community.

To be sure, the important contributions of the analytical method to the task of clarifying meaning are a matter of record. On the negative side, however, there is the still worrisome tendency to confuse clarification with explanation, to conflate tests of the coherence or adequacy of propositions offered as explanations of phenomena with tests as to the nature and reality of the phenomena themselves. These negative effects have been quite pronounced in philosophy of mind, and especially in approaches to the phenomena of consciousness.[33] Consider the difference between a pain as experienced and a pain as rendered public. Even if one agreed that "pain language" replaces grimaces and audible cries, this would have no bearing whatever either on the conscious experience of pain or on an explanation of its causes and effects. It is conceivable that a device could be made in such a way as to change its appearance and make loud sounds when one or another component is destroyed; a device could also be programmed in such a way as successively to substitute verbal outputs for mere sounds. Here, then, we have a device that replaces "pain behavior" with "pain language." We have, in a word, everything but *pain*!

It could be argued that the device in question cannot exhibit either pain behavior or pain language, for it does not have pains as such. Wittgenstein does not deny the fact of pain sensations. Such a fact arises from other facts of a biological and physical nature such that a device of a different nature cannot be said to experience pain merely because it behaves in a certain way. What is at issue in Wittgenstein's analysis is not the fact of pain but the alleged privacy of it, the result of which is that no one can "know" it except oneself and by way of introspection. The famous gambit advanced by Wittgenstein to defeat this privacy claim appears in several places in his work and is illustrated memorably in his "beetle in the box" example.[34] Suppose each person possesses a small box whose contents can be observed only by the one holding a given box. Suppose further that, on examining the contents, each person utters the word "beetle" when asked about the contents. It is clear that "beetle" might mean anything, there being no basis on which to establish that all who use the word are referring to objects of a specific sort. Precisely this problem arises, on Wittgenstein's account, if sensation terms are treated as inviolably private events that, when reported at all, must mean solely what the reporter intends them to mean. Actually, the terms used to express such sensations can *mean* nothing except what is established by usage and local convention. Meaning is socially constructed by participants in the "language game." Were it otherwise, "beetle" could refer to anything at all and, therefore, to nothing definite or defining.

This line of analysis will be considered further in chapter 5, but a few cautionary sentences are warranted here. It is undoubtedly the case that no one knows that another is in pain in the way experienced by the pained person. Groans and cries of distress, perhaps later "replaced" by pain language, give publicity to the sensations. It is not by way of cries of distress or groaning that Smith becomes convinced of his own pains, though these expressions are used by those observing Smith. This is all well and good, though what is unclear is just how it is supposed to undermine dualism or Cartesianism or ghosts in the machine. Descartes' *res cogitans* is not the model of a sensitive or perceptual entity but a *rational* being with the capacity for abstract thought most clearly exemplified by mathematics. The introspective element in Descartes' method is but a form of Socratic testing, a dialogue between one's certainties and one's doubts. Descartes can imagine himself disembodied but only in so far as he can *think* such possibilities. The significant division is not between the public and the private but between the cogitative and the corporeal. Much clarity has been gained by way of Wittgenstein's exposure of certain "grammatical" fallacies, but this very exposure offers a means by which to get further clarity on just what Descartes was attempting.

Tied to the new tradition[35] and its continuing influence, the dominant philosophical method, in the spirit of epistemological narcissism, takes method itself as ultimate, often at the cost of content. Thus, even when an astute critic discloses one side of the problem, concessions are then made that spare the other side. Paul Livingston, for example, while offering enlightened criticism of the tradition of linguistic analysis, then concludes, "our ways of understanding ourselves are inveterately and irreducibly linguistic, that we are humanly fated to determine ourselves through the language that we use."[36]

If this contention is applied to conscious experience, it leads to the conclusion that there is nothing to be explained until language yields something called "understanding," the latter seemingly exhausting the phenomenon. Left unaddressed are whole hosts of conscious experiences for which words cannot be found: experiences of sublimity, of vaguely sensed possibilities, of nascent but targetless desires, of free-floating anxieties, even of odd but perceived objects for which there are no linguistic modes of representation. Nor is it clear in just what sense we "determine ourselves" at all, let alone by way of the language that we use. In *An Inquiry Into the Human Mind* (1764), Thomas Reid argued convincingly that our "artificial language" is grafted onto an innate "natural language" of facial expressions, posture, intonation, etc. (widely shared in the animal economy), by which cooperative behavior is made possible and by which the meaning of the terms of the artificial language becomes settled.[37] On this account, for there to be a linguistic construction of self-determination, the very meaning of the terms thus applied would be grounded ultimately in the culturally unmediated contents of consciousness, or what with reservations I would call *bare consciousness*.

To appreciate how the "problem of consciousness" is to be understood as something novel, one could not do better than Richard Rorty's *Philosophy and the Mirror of Nature*.[38] Not only has the book been immensely influential, but it is a clear and persuasive critique of what its author regards as a veritable tradition within philosophy. Rorty's own philosophical powers have been displayed across a range of important subjects—all the more reason to pay close attention to his views on our present subject.

The title of Rorty's first chapter expresses the orientation of the entire work: "The Invention of the Mind." In these opening pages, Rorty is at pains to establish that, thanks largely to Descartes, there is widespread conviction that mind and matter occupy different ontological slots, neither one being reducible to the other. Owing to this "Cartesian" assumption, we are in the thrall of "Our Glassy

Essence," as he titles part 1. He goes on to argue that such distracting and misleading assumptions are not the gift of an intuitive faculty universally distributed but arise from discursive practices sustained by a defective philosophy. His cure for the malady is provided by, among others, "the three most important philosophers of our century—Wittgenstein, Heidegger, and Dewey."[39] To the extent that each of these writers, in Rorty's judgment, attempted in their later works to disabuse the thinking world of the notion that philosophy is foundational, one has good reason to ask whether the notion had ever been held beyond the rather precious duchy of academic philosophy itself. As noted in the previous chapters, neither Plato nor Aristotle was prepared to credit philosophy with the means by which is established the ultimate nature of "mind" or the means by which it interacts with "body." If, indeed, Rorty's favored trio developed different but comparably effective remedies for a disordered conception of reality, the malady was iatrogenic in the first instance, becoming epidemic as a result of the confining atmosphere of the schoolroom. Ancients aside, the entire thrust of Scottish Common Sense Philosophy in the eighteenth century, led by Thomas Reid, was against the prejudices of the philosophers and, truth be told, Reid and his cohorts did a far better job here than did Rorty's three luminaries. Indeed, some of the best of Wittgenstein shadows whole passages in Reid's *Inquiry*.[40]

I would not wish to reflect lengthily on Martin Heidegger's putative originality in providing "a new set of philosophical categories which would have nothing to do with science."[41] The arguments for the uniqueness of the social were a fixture in the "Romantic rebellion" of the late eighteenth century and thereafter, Heidegger himself acknowledging such figures as Goethe and Schelling in shaping his own views. Rousseau, of course, comes readily to mind as well. Whole chapters could be (and have been) devoted to the influence of Goethe's *Farbenlehre*, Fichte's philosophy of culture, Herder on language, Wundt's anthropological psychology, Windelband's seminal essay on *nomothetic* versus *idiographic* explanation—the list is arm's length.[42] What is especially interesting about Heidegger's rendering is its compatibility with the darker cultural assumptions of National Socialism in Germany during the bad old days. Perhaps it is sufficient here simply to reject the "divided-self dualism" that would have the Heidegger of Nazi party membership intellectually divorced from the Heidegger of philosophy. To accept this is to fail to comprehend either.[43]

In his discussion of the "invention" of mind, Rorty selects as the predictable target of criticism that "Cartesian" mind that is allegedly massless and spaceless and thus nonphysical in the most fundamental ontological sense. He rejects nonspatiality as a criterion of mental states on the grounds that the concept of a state is obscure; moreover, if there are states, they occur in particular entities that must, of course, be spatial.[44] One surely would agree that the concept of a "state" is often deployed in ways that are obscure, but one need not agree that, if there are states, they occur in particular entities that are "spatial." One may record alterations in the information content—the very "entropy"—of the cosmos without having to accept that entropy is "somewhere." Moreover, though accepting the notion of "states" to be ambiguous, in requiring them to be spatial Rorty does not settle the question at issue but begs it. If, indeed, mental states are embodied parts of specific physical particulars then, of course, they are "spatial." Once again, the question at issue is whether all that comes under the ordinary understanding of "mental" is coherently described or explained as an extended part of a particular physical entity. Smith hopes it will not rain on the day of the picnic. There is surely a sense in which the hope, as Smith's, can be found wherever Smith is, but not in the sense that some part or "state" of Smith's body, carefully extirpated, will give us the "hope" without the balance of physical-Smith.

On Rorty's account, we are misled by the nature of intentional objects (objects of thought, hope, desire, etc.) into regarding them as immaterial only because we think of materiality as directly accessible to observation. Thus he says: "Why should we be troubled by

Leibniz's point that if the brain were blown up to the size of a factory, so that we could stroll through it, we should not see thoughts? If we knew enough neural correlations, we shall indeed see thoughts—in the sense that our vision will reveal to us what thoughts the possessor of the brain is having."[45]

The problem gets solved on this understanding by taking intentional objects to be functional states that are, of course, material, though not as discrete, objective, and bounded widgets. The digestive system is in the functional state of metabolizing carbohydrates when its operations conform to the Krebs cycle, though the entire process is continuous, complex, and diverse at the level of individual cells. Granting this much, the promise itself is transparently idle and off the point that Leibniz was pressing. All the "neural correlations" in the world will not yield anything visible, nor is any thought visible as such. As the matter relates to consciousness, there is the added and fatal fact that, even if we were to "see" what Alice is thinking, we would not be "conscious" of *her* thoughts.

As it happens, the concept of an "intentional object" is itself problematic. To say that an "intentional" object is an object of thought raises questions about the status of thought content that has no match in reality; thoughts about dragons and Martians or highways made of clotted cream. Tom Crane has stated the dilemma well: "Either intentional objects are existing objects, in which case it is impossible, contrary to appearances, to think about something which does not exist. Or some intentional objects are nonexistent real objects. But this requires an obscure and implausible metaphysics."[46]

Smith imagines the sun shining for the picnic next Wednesday. As it is not yet next Wednesday, there is no shining sun as of that date. To say that Smith's current "sun" is located, in that it is a material functional state, offers no means by which to distinguish intentional objects from other objects of thought, such as the actually *perceived* sun next Wednesday. The possession of some large enough number of relevant neural correlations might well inform us of Smith's two suns, but as both are objects of thought, the question would linger as to how we might distinguish them ontologically or even psychologically, in light of the fact that at most only one of them is actually in the sky.

Moreover, in both cases, what we would know by some sort of inference (based on neural correlates), about which we might always be wrong, Smith knows in a different way. Suppose we look at all the correlates and declare that Smith hopes it won't rain on the day of the picnic—whereas Smith declares that he hopes it will. Clearly, there must be some basis on which to choose one assertion as an accurate account of Smith's hope. If that basis is empirical, it can be so only insofar as we have performed many correlational studies and have gained ever greater accuracy and reliability in identifying intentional contents. But the project itself must take the asseverations of the given—shall I say the given *res cogitans*?—as the fact with which the neurological correlates must be found and catalogued. At some point in the process, the intentional object can become a "functional state" only in virtue of its being an intentional object. (This is but one reason why we actually *should* be troubled by what Leibniz notes in his *Monadology*.)

Clearly, the otherwise useful resources of linguistic analysis cannot stand as the ultimate method of explanation in the realm of consciousness. The latter, by its nature, is a realm of *contents* and retains this feature whatever causal factors bring it into being or give distinctive character to the contents. What is unique here is not that the contents are unobservable but that they seem to be reserved to one observer uniquely. That observer requires nothing by way of linguistic or semantic resources as the precondition of being conscious and, except in a special sense, not much by way of linguistics in supplying consciousness with content. Again, pain is illustrative but not exhaustive. Patches of color, bursts of sound, hot surfaces—there is more to consciousness than meets the dictionary.

But does all this require two "substances"? Maybe. Does the second substance install Ryle's objectionable ghost in the machinery of the body?[47] I would prefer to turn the question around: what is it in the machine that accounts for the achievements of the ghost? (More on this later.) It should be noted that long before Ryle, other and celebrated figures served up their own systematic misrepresentations of Descartes' argument. I turn to the second of Rorty's three greatest philosophers of the twentieth century, John Dewey. In the essay "Soul and Body," Dewey offers what he regards as an explicit corrective for the "Cartesian" error. I will juxtapose passages from Dewey and Descartes so that one might assess the progress made in the three centuries separating the two philosophers. In "Soul and Body," Dewey writes,

> The psychical is *homogeneously* related to the physiological. Whatever is the relation of the psychical to the neural, it is related in the same manner to all parts of the neural. The brain is no more the organ of mind than the spinal cord, the spinal cord no more than the peripheral endings of the nerve fibres. . . . Now this gives us but one alternative: either there is absolutely no connection between the body and soul at any point whatever, or else the soul is, through the nerves, present to all the body. This means that the psychical is immanent in the physical. To deny this is to go back to the Cartesian position, and make a miracle of the whole matter.[48]

And from Descartes' *The Passions of the Soul*: "We need to recognize that the soul is really joined to the whole body, and that we cannot properly say that it exists in any one part of the body to the exclusion of the others. For the body is a unity which is in a sense indivisible."[49]

On this point, at any rate, there is no difference whatever between Descartes' and Dewey's positions, though Dewey, as early as 1886, is already content to dismiss a "Cartesian" philosophy of mind without pausing to understand it in its fullness. The point here is not that Descartes and Dewey are in agreement as regards mental life; they are not. The point is that "Cartesian," if it is to be an informing modifier, turns out to be a subtle, deeply considered, and carefully developed position rather than an unwitting howler committed by a philosopher whose powers of analyticity fall below those enjoyed by the indifferent undergraduate. Whether Descartes had in mind just what Dewey is getting at or, for that matter, what more recent writers have proposed[50] is not the point. The point is rather that Descartes is not the comic hero of "Cartesianism" but a theorist strongly inclined toward philosophical materialism while, at the same time, having to acknowledge the conundrums arising from attempts to apply that materialism to the domain of consciousness.

A word more on Dewey as Rorty would have him understood, or perhaps misunderstood. Dewey is routinely treated as the defender of the new pragmatism to which Rorty himself is committed. The ruling maxim is that total democratization of thought and theory grants unrestricted access to the public square. Only through the practical working out of the welter of possibilities might one or another script be vindicated. Nothing is foundational as such; all is relative to context and to local aspirations, local values. If, however, it is this version of neopragmatism that Rorty takes to be one of the signal contributions to philosophy, he cannot offer Dewey as its harbinger. When it mattered most in the actual public square, Dewey proved to be rather more the foundationalist than the radical democratizer. I refer here to the controversial question of communists in the 1930s retaining membership in the New York teachers' union. It was Dewey who led the charge against the group, recognizing the incompatibility between their cause and the very purposes of a liberal education.[51] A separate book would be required to clarify just how eccentric is the juxtaposition of Dewey and Heidegger on a short list of the century's philosophical greats. I note Dewey's rejection of the communists as a guide if we would ponder how he might have dealt with Heidegger's own party affiliations.

Another "Cartesian" core issue in philosophy of mind is that of "mental causation." In a nutshell, the problem is this: just in case there is an ontologically valid category of the "mental," the properties of which are, among other features, nonphysical, how is it that mental events or processes causally bring about physical events such as raising an arm or voting for a flat tax? As Robert Audi notes, "the causal power of the mental is at once pervasively presupposed in common-sense thinking and widely disputed among philosophers."[52]

There is something of a tradition in metaphysics according to which for "a" to be the "cause" of "b," it is required that "a" and "b" have something in common, if only spatial contiguity. But if "a" is mental—understood now as nonspatial and massless—and if "b" is physical—e.g., raising an arm—then how can "a" bring about "b"? Donald Davidson, setting out to oppose the Cartesian dualism seemingly required to remove the conundrum, advanced a theory of *anomalous monism* that endorses the conceptual but not the ontological distinction between the mental and the physical. At base, everything is physical, but it will ever be necessary to consider the mental in its own terms.[53] What is "anomalous" is that an in-principle physical process must nonetheless be conceptualized in mentalistic terms.

There are three principles advanced by Davidson in his attempt to establish the ontology of the mental as physical while showing that the mental is nonetheless not *nomologically* reducible to the physical.[54] The first is the *Principle of Causal Interaction*, which regards at least some physical events to be the effect of mental events. If, in fact, the *Bismarck* was sunk, this came about owing to the plans, desires, and beliefs that gave rise to those actions that led to the sinking of the *Bismarck*. Davidson takes this to be uncontroversial, though note that what is not explained here is just the problem of how plans, desires, and beliefs get muscles to contract. The second principle is dubbed the *Principle of the Nomological Character of Causality*, and expresses the thesis that causes and effects are related by strict laws. This then leads to the third principle, the *Principle of the Anomalism of the Mental*, which refers to Davidson's contention that "there are no strict deterministic laws on the basis of which mental events can be predicted and explained."[55] This claim is based on the sound recognition that events qualifying as "mental" are so utterly contextual, related in so many and such various ways to their antecedents, and so susceptible to rapid movements in new and unanticipated directions that any scheme of prediction is doomed. In a word, "mental and physical predicates are not made for each other."[56] No surprise then, for all the effort expended on reductive strategies, the mental cannot be absorbed into that explanatory and descriptive space within which occurs all that is physical. At the end of the ontologist's day, there are only physical entities, but for one who would seek to account for what arises from the mental, this truth is finally of little assistance.

Davidson's defense of a monistic ontology is subtle, powerful, and influential. Common sense alone forces the recognition that at least some mental events are causally related to physical events, whether the latter are as striking as sinking the *Bismarck* or as unremarkable as turning a page to get to the conclusion of the story. So, too, is it obvious that at least some singular causal relations are lawful. How else did men get to the moon and back? Is such lawfulness absent at the level of psychophysical relations? Consider (a): Smith "wills" to raise his arm and does so, this being evidence of the causal power of the mental in relation to the physical. If such an antecedent-consequent relation obtains, it is because a strict law is operative. But there is no "strict" law that enables prediction and explanation of the hopes, beliefs, desires, perceptions, etc. of fully developed, acculturated, and active human beings.[57]

I must contend, however, that Davidson's analysis does not achieve his ends. In his adoption of such a model of analysis and explanation, the physicalist conclusions at issue become virtually foreordained. The very language of "covering laws" harkens to a model of explanation advanced by the logical positivists and long accepted within the scientific community. However, it is just this model that is under consideration and just such a database that would be offered as a corrective. When Davidson carefully restricts his thesis to events that reflect his *Principle of Causal Interaction*, there would seem to be no room for any but a physicalistic ontology. At least one arguable reason that there is no "strict" law that joins the psychic and the physical is that mental events are not the result of physical-physiological processes but of other *mental* events. Surely, if Jack "wills" to open the door by turning the key, and if Jack thereupon opens the door by turning the key, there is some relationship between the volition and the act of opening the door. But to require that it be a *causal* relationship is to beg the question. More on this general issue will be developed in chapter 7. At this point, it is sufficient to acknowledge the alarm that might arise from suggesting relationships other than causal ones, especially in such transparent occurrences as opening a door with a key. Is this not, however, what has engaged philosophers of mind for decades? There is a doubtless spatiotemporal relationship between the desire to enter the room and actions that result in the door to that room being unlocked through the use of a key. There is a doubtless mental-physical relationship between the desire to enter the room and the motor activity associated with the competent deployment of the key. Needless to say, the conundrum evaporates if we can show that the desire as such is no different from any other physical force and thus overcomes the inertia of the arm and hand. But desire just won't conform here. Davidson is right, of course, when he says that mental terms and physical terms are not meant for each other. The "problem of consciousness" is in part the problem of finding out why not.

We might return yet again to events Davidson agrees to call "mental" and see how they operate within settings calling for decisions and judgments. Consider the following transmission over a communications channel, such that, if indeed there is a cat in the tree, the recipient of the message will deploy a rescue team:

THE C _ T IS IN T _ E T _ _ E

What we see in the message is a missing *a* in the first noun, a missing *h* in the article, and a missing *r* and *e* in the final noun. Nonetheless, owing to the redundancy of these letters, the recipient of the transmission records that, indeed, THE CAT IS IN THE TREE. On the strength of this, the fire department is sent to the scene. Now, what is it that "causally" brought about the rescue? Only in an odd sense would we consider something "physical" as such. What "caused" the fire engines to leave the station was a packet of *information* that was causally efficacious but surely not an originating source of physical energy. What moved the rescue team was the *information* that supplied a *reason* for action, not a physical *sound* that served as a cause. Reasons are "causal" but not *physically* causal. Information is causal but not *physically* causal. What we must get used to is that there are some causal sequences that are not physical at base. This is different from getting used to what we already accept; viz., that mental events are not usefully translated into the language of the physical sciences. Where facts render a theory untenable, we must not jettison the facts!

Davidson has long been on the front line of important philosophers seeking to rid the world of thought of Cartesianism. It is clear, however, that much of the anti-Cartesian labors of the day recover what is explicit in Descartes' own work and that the more weighty criticisms were advanced by Descartes' celebrated contemporaries.[58] "Anti-Cartesianism" is now largely "code" for a defense of a physicalism no more credible and no more coherent than what Descartes had attempted to defend. His was very much a form of "cognitive neuroscience" that would be surprisingly at home in today's major centers of thought on these matters. In the next chapter, we will examine approaches to the problem of consciousness based on the notion of "higher-order thought" to see if, perhaps, "Cartesianism" will appear at all primitive by comparison.

4

HIGHER-ORDER THOUGHT
A Machine in the Ghost

There is a range of theories about consciousness judged to be "leading edge" and influential in contemporary cognitive neuroscience. They all have much in common, not the least of which are unintended "Cartesian" features—not to mention an often spirited "refutation" of Cartesianism. Yet another common feature is the implicit (when not explicit) endorsement of the brain sciences as foundational for the theories themselves. What is not questioned is the metaphysical commitment to one or another form of *physicalism*. If, then, the elusive and even wraithlike nature of consciousness is to be explained in a manner compatible with this commitment, ways must be found by which to install some sort of machine in the ghost.

One such theory is referred to as HOT, the acronym for *higher-order thought*.[1] Developed under this heading is the thesis that there is an important distinction between what one is conscious of and an awareness that there is a conscious process by which to reach what is currently not available to consciousness. It will be useful to consider this theory at length because, as I say, it is illustrative of a variety of conjectures widely shared among today's specialists in "cognitive neuroscience." Moreover, it also illustrates something of a style of expression that conveys notions of far greater specificity than is actually incorporated into the theories themselves.

HOT is a *metacognitive* theory regarding the access one has to one's own cognitive states. How to illustrate this? Tip-of-the-tongue phenomena are of service here. One may consciously strive but fail to recall the name of the author of *Brideshead Revisited*, all the while aware that the name is potentially reachable by some sort of mental process. One might even employ an alphabetical heuristic: A . . . , B . . . , etc. Just in case, on reaching "W," one now recalls *Waugh*, the previously nonconscious content is said to have entered into consciousness. One might then have greater confidence in the heuristic and employ it often when in a metacognitive state. At bottom, the theory requires of any state that is "conscious" some higher-order thought associated with it. Thus, the perception of a round red patch would not count as a conscious state unless and until some higher-order thought (e.g., "apple") was associated with it. The lower-order state on this account is just a physical state of the brain. Consciousness arises when some higher-order thought or belief or attitude is tied to it.

There seems to be a question-begging problem lurking in the wings of this theory. If one aim of the theory is to naturalize or demystify consciousness, it is scarcely economical to do so by requiring complex thoughts and attitudes as necessary ingredients. Nor is the source of the latter at all clear. Alvin Goldman draws attention to the problem this way: "How could possession of a meta-state confer subjectivity or feeling on a lower-order state that did not otherwise possess it? Why would being an intentional object or referent of a meta-state confer consciousness on a first-order state? . . . Why should a first-order psychological state become conscious simply by having a belief about it?"[2]

Without further examining this major feature of HOT-type theories, it is clear that they are "Cartesian" at least at the level of establishing differences between perceptual and cognitive processes and between mere sensation and actual reflection on one's own cognitive states. Advocates of HOT might wish not to be dubbed "Cartesian," but the conclusions they have reached arise from the very mode of analysis Descartes advanced in his *Discourse*. To say that a mental state is "conscious" only to the extent that one has some sort of "higher-order" thought about it is to say that there must be a *res cogitans* reflecting on experiential content and distinguishable from an entity possessed of no more than perceptual, motivational, and emotional states. Descartes leaves no doubt about the limitations of all nonhuman creatures who, lacking rationality, may outperform human beings in many ways but in none requiring thought as such. As he insists in his *Discourse*, animals even outstripping human beings in certain performances "do not act from knowledge, but only from the disposition of their organs."[3]

HOT theorists wish to contrast their position with "Cartesianism" by emphasizing the frequency with which one is guided not by the contents of consciousness but by the contents of other states to which the person does not have conscious access. Descartes, of course, routinely acknowledged the fallibility of cognition. His very "method of doubt" is predicated on the well-known distortions to which introspection is heir. The main point, however, is one the HOT theorists share with Descartes: it is by way of *thought* exercised on the contents of experience or of revived experience that one is properly regarded as being in a conscious state. Descartes was convinced that such thought cannot be achieved by matter whereas HOT theorists seem convinced that it can. What Descartes finds when entering the "machine" is *machinery*, and this is all that the HOT theorists find as well. They conclude from this that conscious states and thought somehow arise from operation of the machinery, as Descartes concluded that such was not possible. As noted already, the mere stipulation that *physics is complete* offers no compelling reason to choose HOT over Descartes' thesis. For all the progress (and there has been great progress) in the brain sciences since the time of Descartes, these alternative positions retain their credibility. Descartes' has the edge to the extent that it requires nothing of matter beyond what modern science is able to discover.

On this latter point, note that the HOT theory itself is shrouded in mysterious assumptions and turns of phrase. As noted earlier, there is a fairly liberal use of the concept of "states" and "processes" in this entire domain. It is utterly unclear that one is aware of being in a "conscious state." One is simply *aware*. As all the evidence on which such turns of phrase are based is drawn from first-person accounts, it is not immodest of the author to submit his small share of data to the theory: I am *not* aware of being in a "conscious state," for my awareness of any state at all presupposes that I am conscious. This includes not only belief about being in a given state but also states of happiness, anxiety, confusion, whimsicality, etc. The exception to all this is when I am so completely absorbed by a difficult task that the demands of the task seem to override self-reflecting thought. It is the *doing* that dominates, not an awareness of the doing. The same may be said of states arising from significant aesthetic experiences where the sublimity of the work creates something of a reverie to which the property of "conscious of" seems uniquely unsuited. I say all this without at this point being scrupulous as to the notion of "states" themselves.

Referring back to the example of tip-of-the-tongue phenomena, I add this report: If I am struggling to recall the name of the author of *Brideshead Revisited* and begin to try various heuristics, I am, of course, aware that I'm using a mnemonic technique—after all, I *chose* it. But I am not "judging," nor am I entering a metacognitive "state." As the only evidence available to others of what I'm doing will come from me, what I will tell them is that it is sometimes useful, when one cannot retrieve a name, to start the search for it by going through the alphabet. It's a bit of a leap to move from this ancient fact of folk psychology to a general theory of consciousness, unless the general theory is just an unnecessarily technical restatement of folk psychology.

The Rosenthal version of HOT restricts conscious states to

those mental states about which one has "higher-order thought," the thought itself somehow caused or brought about by the mental state. The theory therefore includes as "mental states" those that are not themselves conscious but that have some means by which to trigger higher-order thought about *themselves*. The *Brideshead* example is again useful: Somewhere in the ongoing but nonconscious mental states is the putatively coded representation of "Waugh." As I engage in higher-order thought about "the author of *Brideshead*," the representations in nonconscious mental states sustain conscious states that finally retrieve *Waugh*. Granting that there is nothing in the account that is provably wrong—or provably right—the natural question would have to do with just what we now know about the nature of consciousness that was obscure or misguided until the HOT theory was propounded.

Surely one of the more obvious problems with theories of this sort is the interchanging of "being conscious" and "being conscious of," as if the two were either synonymous or mutually entailing. Charles Siewert advances an informing criticism of this, noting that the reference of "conscious of" need not be an intentional object. Thus, "clearly the 'of' in 'thought of' is often the 'of' of intentionality, but we would not suppose that by 'a thought of some merit' we must mean something like: a thought about ('directed at') some type of merit."[4]

Apart from such important considerations, there are theoretical controversies that surround versions of the HOT theory, some favoring a dispositional and others a nondispositional variant. We might use the shorthand notation "HOT-d" and "HOT-nd" to identify these types. Peter Carruthers, an advocate of HOT-d, proposes a system in which states that are "conscious" in the phenomenal sense have intentional content—they are *about* something—but are causally brought about by mental states devoid of such conceptual content but integral to the HOT associated with the nonconceptual content. At work on this account might be some sort of buffer-store of which one has no phenomenal consciousness but whose content is available to consciousness just in case the relevant triggering functions occur. It is in virtue of the contents of the buffer that one is *disposed* to higher-order thought of a specific nature. HOT-nd versions are similar with only arguably significant differences. On this account, a mental state is phenomenally conscious to the extent that its nonconceptual content is the object of HOT, the nonconceptual content causing the thought about it.[5]

Carruthers, as with other HOT theorists, seeks to "naturalize" consciousness by translating its contents into representations.[6] Accounts of this sort, widely adopted within contemporary cognitive science, raise more problems than they settle. The very concept of a "representation" is generally neglected or is explicated by way of arguable analogies. It is a strength of direct realism that it requires no such spurious theory of "representations." What is seen is what is before the percipient, not a representation of it. As I look at the words appearing on my computer monitor, I do not see a "representation" of the words but the words themselves. The major weakness of representationalism is that there is no evidence for it, nor is there a coherent account of just how a neural process or event becomes translated or transformed into a depiction of reality.[7]

HOT theories are now presented in book-length treatments and are featured at special conferences. It must be acknowledged, however, that they seem to contribute very little to our understanding of the nature of consciousness and that phenomenal quality that stands as the putative *problem* of consciousness. What the theories assert is that for mental states to qualify as conscious, it is necessary that there be some sort of thought or belief *about* the contents of the perception or experience or desire. There must be thought *about* some sort of nonconceptual *something* lodged in some other state or process or even place. It is unclear as to what makes the such thought "higher order," for the term, as routinely used, stands for no more than the allegedly required "higher functions" of the brain. However, the crux of the mind/body issue is the very question of how material processes in a biological system bring about the phenomenal features of mental life.

To answer with a theory that presupposes the efficacy of just such processes is once more unhelpful.

And then, what is a *lower-order thought*—a LOT? And what is the metric or threshold point dividing HOTs and LOTs? Does one reach the former by adding up the latter? Are nonhuman animals possessed of HOTs and, if not, does this relegate them to the realms of the nonconscious? I will return to these questions later in the chapter and consider them in a nonrhetorical fashion. First, some attention should be given to the sorts of findings that support the notion of nonconscious mental states that have causal roles in conscious mental states. *Blindsight* is one such finding, often advanced as an illustration of a nonconscious mental state that causally brings about the content of the conscious mental state of thinking about or reflecting on it.[8] The "blind" patient, victim of a neurological disease, consistently denies conscious awareness of light but nonetheless makes responses clearly indicating the position of the *unseen* light. The patient who does not "see" a visual stimulus nonetheless points to its "guessed at" location with greater accuracy than would be possible on a chance basis. The conclusion is that a nonconscious mental state is causally effective in providing content for a genuinely conscious mental state.

Not considered within this theoretical framework is the reasonable criticism that the nonconscious states are, for that reason, not *mental* as such. A limb separated from the body by accident will display neuromuscular integration for many hours under proper conditions but, from the fact that stimuli elicit predictable responses of which the amputee is not conscious, no one would argue that there is evidence of a disembodied "mental" state. What sort of state must a state be to be *mental*? If it must be causally effective in arousing thoughts about its "content," we then have the problem of establishing what the "content" of a "state" is, other than its sheer physicality. The reply that the content is a "representation" is, to say no more, not helpful.

It is worth recalling that William James considered the HOT line of reasoning in his discussion of "mind dust" theories in his *Principles of Psychology*. Those James referred to as "mind dust" theorists advanced the view that elementary "mental" events were associated into larger clusters, that "consciousness" arises by gradations from the nonconscious primordial, etc. Opposing various arguments favoring such theories, James includes this:

> *Third Proof*. Thinking of A, we presently find ourselves thinking of C. Now B is the natural logical link between A and C, but we have no consciousness of having thought of B. It must have been in our mind "*un*consciously," and in that state affected the sequence of our ideas.
>
> *Reply*. Here again we have a choice between more plausible explanations. Either B was consciously there, but the next instant forgotten, or its *brain-tract* alone was adequate to do the whole work of coupling A with C, without the idea B being aroused at all, whether consciously or "unconsciously."[9]

In other words, the "idea B" may well not count as an idea; it may be some intermediary state of the nervous system and thus not "mental" in the required sense. Or it might just be found and lost so quickly as to leave no conscious trace or record. It is doubtful that James would be patient with the notion of brain states having conscious or unconscious "content." Referring to "states" having "contents" is not to explain either the mental or, for that matter, the nonmental. Consider as an unarguably nonmental set of facts the breakdown of a molecule of photopigment in one retinal cone cell. The chemical change is causally associated with the generation of a weak d.c. signal in bipolar cells functionally connected to the cone receptor. These changes, when of sufficient amplitude, lead to electrical discharges in the axons of retinal ganglion cells whose axons compose the optic nerve. To the extent that optic nerve discharges causally bring about visual experiences, it would be informing to say that a nonmental event is followed by a visual experience under certain conditions. Put tersely, an electrochemical event reliably elicits the experience of light. Far less informing is terminology that would have changes in the pigment chemistry referred to as examples of *nonmental content*, for there is no "content" at all. Rather, there are identifiable and recordable physicochemical events that are not "like" anything other than still other physicochemical events. They are not "representations" or codes or premental or any such thing, for they can just as well take place in a glass beaker.

More than thirty years ago, Thomas Nagel put forth the productive question, considered in the previous chapter, "what is it like to be a bat?"[10] There must be something it is "like" to be a bat. There is something it is like to be a baby. But there isn't anything it is "like," presumably, to be a photopigment or a stone or a river. Following Nagel, David Chalmers has stated the "hard" problem of consciousness this way:

> The hard problem of consciousness is the problem of experience. Human beings have subjective experience: there is something it is like to be them. We can say that a being is conscious in this sense—or is phenomenally conscious, as it is sometimes put—when there is something it is like to be that being. A mental state is conscious when there is something it is like to be in that state.[11]

This way of distinguishing consciousness from something (everything) else is not free of problems. One may agree that the severely mentally disturbed patient, convinced that he lost the battle of Waterloo, is in a conscious mental state owing to the fact that there is something it is like to be in that state. Less clear is whether the state in question reaches what it is like to be Napoleon or what it is like to be psychotic. It is quite obvious that the patient's consciousness cannot reach what it is like to be Napoleon or what it is like to be psychotic. To say instead that the contents of consciousness here establish what it is like to be an *experient* is as informing as the definition of a sleeping pill as one possessing the *vis dormativa*.

There may be some obscure, arguable dividing line within the biosphere on one side of which creatures are such that it isn't anything to be "like" them. The phrasing is again rather confusing. To say there is nothing that it is like to be a stone calls for the reply, "except a stone." *What is it like to be "x"?* is more akin to musing than to questioning. To say, then, that a mental state is conscious when there is something it is like to be in that state is a cumbersome way of defining consciousness in terms of experience, but now we have the hard problem of experience to contend with. Progress is not won by changing the order of words to: "The hard problem of experience is the problem of consciousness." Moreover, it is counterintuitive to theorize that only with HOT is it possible to be "like" a creature with consciousness. It is merely argumentative to dismiss every nonhuman form of problem solving, behavioral adaptation, and expression of feeling as empty of "thought," let alone HOT. Luis Bermudez has convincingly challenged the allegedly conceptual connection between language and thought.[12] Though denying nonlinguistic creatures the capacity for inference, he records any number of illustrations of their "thinking," at least as the activity earns that designation in human communities.

Of course, the concept of "thought" is itself not precisely developed and probably cannot be. Direct evidence for it is personal, introspectively reached, and graded not in terms of "higher," "lower," or "first-order" but with such terms as significant, perplexing, fanciful, pressing, idle, frightening, etc. All such descriptions are meaningful under those conditions of life characteristic of human life. To be a bat or cat or turtledove is to face other conditions and thus to enter into engagements with them likely to be species specific. Fish will not "discover" water, for they will not have life once removed from it. Were we aquatic by nature, there would be radically different HOTs about H_2O, I suspect. The point, of course, is that "experience" itself does not lend itself to facile transspecies comparison. It is arguable enough when contending to know the experiences of another person without daring to legislate on just what nonhuman animals record and retain from encounters with a world of stimulation. Perhaps for human beings the mode of consciousness calls for the addition of HOT for that mode to be engaged and usefully deployed. Who would speculate on just what the necessary and sufficient ingredients would be for "consciousness" as it may be occurring in the hamster and the duck? In human vision, maximum sensitivity to color falls at electromagnetic wavelengths of about 555 millimicrons. The peak for the honeybee is in the ultraviolet region of the spectrum where human vision is blind. What the bee "sees" is not available to human observers and therefore is not a subject on which they can speak with any authority. In the matter of phenomenal experiences, authority is a first-person matter, all generalities validated by summing across the first-person accounts. This is how we know that, for example, the dichromat suffers some sort of deficiency when declaring the red apple to be gray. One can specify quite accurately the narrow range of wavelengths that will result in the utterance "red" by all color-normal percipients. This still leaves unanswered whether any two of them see "the same" red. If whether two different persons see the same red remains a philosophically coherent question, then what the honeybee sees must be ever more daunting, for on this question there is no relevant baseline from which we might make generalizations.

The HOT theorists have attempted to give a tighter and workable character to the concept of consciousness, to distinguish it from attention or perception or, understood in a certain way, even awareness. The criterion adopted to achieve this end is, however, simply unconvincing and counterintuitive. One is conscious, on this account, only to the extent that one is simultaneously thinking that one is in a given mental state. A conscious perception requires a cognitive awareness that perception is, indeed, the process taking place. As noted, this criterion does not seem to be commonly met by percipients.

Added to this is new HOT wine is the old "limen of consciousness" bottles that nourished the development of psychophysics. Thus:

1. There is some stimulus magnitude so weak as to be imperceptible.

2. Adding a unit of magnitude to this will not succeed in triggering a perceptual outcome until some critical (threshold) magnitude is reached.

3. Nonetheless, the sense organs are in some way affected by the subthreshold stimuli, now working at an *unconscious* level.

4. It is the accumulation of the unconscious events that finally engage the conscious awareness of the percipient.

This was all developed at length by Johann Friederich Herbart, Kant's successor at Königsberg, in his 1816 *Lehrbuch der Psychologie* to such an extent that Fechner regarded Herbart as nearly the founder of psychophysics.[13] I note this not to establish theoretical priority but to note that even two hundred years ago the general idea was based on a common-sense understanding of what it takes for events to enter consciousness. Herbart's analysis includes instances in which thoughts are in such conflict as to cancel each other while still disposing the person to think still other thoughts.

In the end, HOT theories tell us that we are conscious to the extent that we are aware of the states we are in, whether perceptual or volitional or credulous. This is obvious. To expand the obvious in order to illustrate the defects of an alternative theory is of value, but only to the extent that the alternative is itself credible and useful. The alternatives to "higher-order" notions are, alas, "first-order" notions as to how the external world comes to be an object of consciousness. Fred Dretske, eager to "naturalize the mind," is among those who have defended the "first-order" account, though the theory, so called, would seem to have little that would warrant much by way of a defense.[14] It is an unwitting rehash of Herbart-type notions: some sort of "fine-grained" or "analog" events not themselves subject to conscious regard causally bring about consciousness. What seems so immaterial about the latter is that it arises from processes so fine grained as to be beyond the reach of the phenomenal. Of course, the obvious question to ask is that, now that we know that the experience of the yellow rose arises from events that begin with changes in the pigment chemistry of the retina and come to include unconscious events in neurons, etc., just what is it about our *phenomenal experience* that is now understood? This is no way to "naturalize" a mind!

Common sense and the daily experience of legions of men, women, and children are not to be charged with "mystery mongering" by theorists who "solve" the problem of consciousness by confining it to events of the sort dominating the Monday seminar. There is a *philosophical* or, perhaps, *metaphysical* problem of consciousness, but it is theory driven by ontological and explanatory commitments arising from modes of analysis that are themselves problematic.

I return to the rhetorical questions previously raised: What is a *lower-order thought*—a LOT—and what is the metric or threshold point dividing HOTs and LOTs? Does one reach the former by adding up the latter? Are nonhuman animals possessed of HOTs and, if not, does this relegate them to the class of the nonconscious? As discussed in the first chapter, Aristotle was of the view that human rationality conferred features on mental life not found among other animals. Through rational power, the adult and normal human being is able to frame beliefs and hold convictions about that which has been delivered via the senses. As the nonhuman animal's commerce with the external world is at the level of biology, all representations (*phantasmata*) are akin to the formation of material impressions on a wax tablet. There is no "truth" or "falsehood" here, for these terms apply to beliefs, of which, on Aristotle's account, nonhuman animals have none.

If this is all that HOT theories are claiming, then they do not really address the problem of consciousness. To be "conscious" of a stimulus, a creature need not have beliefs or convictions. If the addition of belief raises the level of mentation to that of HOT, then the theory is just Aristotle's and much ink can be saved, except for the problem of consciousness itself. Needless to say, still other difficulties arise once a theory of consciousness excludes all but one member of the animal kingdom. Such a theory must put on notice everything within evolutionary psychology that is predicated on the principle of continuity, for such a psychology must find primordial forms of consciousness at least among the near relatives of *Homo sapiens* (*sapiens*!). That such relatives have, as it were, "less" of it doesn't work very well, as consciousness is usually taken to be a binary state. One either has it or not. There are degrees of it, to be sure, as we find in "twilight" states or light anesthesia, but it would be odd to characterize the full range of behavioral achievements displayed by, say, chimpanzees as the product of lightly anesthetized states.[15]

There appears to be a property kinship between the domain of consciousness and that of morality, though one must hurriedly explain oneself lest there be a whiff of the mystical. More than a century ago, G. E. Moore argued that moral properties are unique and entirely beyond the range of things and events understood to be natural. Moral properties on this account are nonnatural.[16] The argument roughly is that natural things and events can be exhaustively described in the language of the natural sciences. If moral ascriptions were drawn from the same ontological domain, they would stand in synonymous relation to natural-physical ascriptions. Accordingly, replacing moral ascriptions with naturalistic ones should be as "analytic" as, say, replacing "unmarried male" with "bachelor." Referring to this as the "open question" argument, David Brink provides this useful summary: "Consider any moral predicate 'M' and any non-moral predicate 'N'. If 'M' and 'N' mean the same thing, then it ought to be an analytic truth that M-things are N, just as it is an analytic truth that M-things are M. . . . It is not possible to doubt that M-things are M—'Is this M-thing M?' is always a closed question. However . . . 'Is this N-thing M?' is always an open question."[17]

Comparably, one might raise the question, "Is this B-state B?", where "B" is the aggregate brain activity during a given sampling epoch. That a brain state is a brain state is a closed question. "Is this C-state B?", where "C" is the declared first-person report of awareness or consciousness, remains an open question. It is open in more than the logical sense. During a given sampling epoch, aggregate brain activity is just that and, as such, will be perfectly correlated with all sorts of subdural events and processes, including those that are digestive, respiratory, lymphatic, etc. And the same would be the case were "B" restricted to aggregate cortical activity. Sudden changes in "B" reliably affect "C" and, if only for the sake of argument, might be accepted as the causal states absent which "C-effects" would not occur. However, a causal account of how a given C-state is brought about does not "explain" C-states or solve the "C" problem. Indeed, the ease with which persons volitionally divide, shift, and focus C-states raises the possibility that they bring about the very brain states causally required for certain cognitive achievements. Thus might the "ghost" make the machinery do its bidding. Physics may not be complete after all, if the more complete physics might have to stretch itself to accommodate what, by current lights, would seem to be as strange as, say, multidimensional space. This, of course, is now part of the common thought in theoretical physics, though the "space" in question would seem to be forever beyond the perceptual reach of creatures such as ourselves.

In light of these difficulties, one is reminded of William James's taunting question and the answer he gave in "Does 'Consciousness' Exist?"[18] Endorsing a version of the open-question argument, he begins his famous essay with the observation that " 'thoughts' and 'things' are names for two sorts of object, which common sense will always find contrasted and will always practically oppose to each other."[19]

As a word, "consciousness" for James is either jejune or it refers to a function that James takes to be *knowing*. With consciousness, reality contains not only things but things that "get reported," as he says, though not by way of some "aboriginal stuff" utterly different from the material things of the world. So far, this seems headed toward eliminativism, but then James makes clear just what unitary kind of "stuff" constitutes reality:

> My thesis is that if we start with the supposition that there is only one primal stuff or material in the world, a stuff of which everything is composed, and if we call that stuff "pure experience," the knowing can easily be explained as a particular sort of relation towards one another into which portions of pure experience may enter. The relation itself is a part of pure experience; one of its "terms" becomes the subject or bearer of the knowledge, the knower, the other becomes the object known.[20]

This monistic ontology was taken by James as the least contradictory, the least mysterious reckoning of reality as it may ever be known or understood. The entire domain of the knowable, whether practical or theoretical or speculative, is the *here and there* in which pure experience is manifest. The known and the knower are inseparable when the function of "knowing" is in operation. Attempts to reduce one to the other, to collapse knowing to some material thing, is to fail to comprehend that both sides of the dyadic relation fall within "pure experience" itself. On this same account, it is also a misapprehension to think of the knower as somehow outside the relation. What is unique about knowing—about "awareness"—is that the fact that is in principle knowable becomes known: "The peculiarity (of consciousness) is that *fact comes to light* in it, that *awareness of content* takes place. Consciousness as such is entirely impersonal—'self' and its activities belong to the content. To say that I am self-conscious, or conscious of putting forth volition, means only that certain contents, for which 'self' and 'effort of will' are the names, are not without witness as they occur."[21]

This will not do, of course, for an infinite regress follows the reference to the required "witness." If the overall content of experience includes "self" and its volitionally expended effort *plus* a witness, then we have an ontology that includes not just pure experience but the occasional witness when that experience includes "self" and "volition." I will return to this issue in the next chapter, which is devoted to self-consciousness. Anticipating the argument to be developed there, I should say that the sense in which consciousness is owned is different from the sense in which something is "witnessed." Jack and Jill, we stipulate, have identical brains in all respects except location. Jack's is inside Jack's cranium, Jill's within Jill's cranium. Moreover, their experiences have been relevantly identical, as are the apples both of them are examining. Jack and Jill both report that they are "witnessing" apples, these now items in their conscious experience. We might go so far as to say that, in some strained sense, even the experiences are identical, save for occurring in different places. However, Jack's consciousness is Jack's, full stop. His belief that Jill is conscious is inferential. His awareness that he is aware is not. Whatever else might be said about consciousness, it is assuredly not "impersonal."

Nonetheless, James's emphasis on "pure experience" as a monistic and "dual-aspect" ontology equally compatible with the reality of thought and the reality of things is a plausible alternative to the dualistic ontology embraced by Descartes. This is not to say that James's thesis is metaphysically or logically or, for that matter, "scientifically" sounder or more compelling, only that it does no violence to common sense, to language, and to the facts as we now possess them.

I return now, but briefly, to the "open question" argument advanced by Moore in defense of his conclusion that moral properties are nonnatural. Having applied the same argument to the claim that mental properties can be matched to physical (brain) states, I must include reservations regarding the success of the argument in either context. A defender of moral realism in other works,[22] I forfeit the right to adopt Moore's position on moral properties and, by extension, the same position applied to consciousness. At issue, of course, is just what is taken to be "natural," which, needless to say, returns us to the "physics-as-complete" thesis discussed previously. It is sufficient to consider the enlargement of the periodic table over the most recent hundred years to abandon ontological complacency. With six distinct leptons and six different "flavors" of quarks now on record, there is ample evidence that even our current (not to mention our traditional) ontologies of the "natural" are incomplete.

Once ignorance is acknowledged, a respectful silence should ensue. To grant that there may be ontological surprises in store for us is not to be able to say anything about whether such surprises will simplify, complicate, or pass over the problem of consciousness, at least to the extent that the problem continues to be regarded as a challenge to ontological parsimony. A two-stuff reality is vexing enough, and ever more so when the two are supposed to be in some sort of causal network or relationship. That this all seems "unnatural" may, however, have more to do with the current state of our knowledge of the natural than with anything ultimately queer about the mental. It is also useful here to caution against inadvertent question begging. Electrical charges are "natural" but have no mass and thus are not "material." Quarks do have mass but it is not directly measurable, as the quarks are captured in a manner that prevents this. The broader sense of something being natural is that it is not artificial—it is not an *artifact* made or manufactured. Surely consciousness is not artificial—or is it? This question, too, must be considered. Whatever states of wakeful attention and thought we have in mind when applying the term, these were surely available to any and every competent and adult being qualifying as a human being.

As noted (repeatedly), there is a strong tendency to absorb all that is "natural" into the domain of the physical, but this is an absorption that fails at many levels. It was Aristotle's contention that social and political modes of organization are entirely *natural*, but it would be nonsensical to attempt to translate the political and social practices of a community into a set of physics equations. The respects in which social beings are social are different from the respects in which they are also physical. The respects in which they are social is, however, as "natural" as are the respects in which they conjugate proteins. But might they be, along with consciousness itself, no more than social? Here is a taxing question to be considered next.

5

SELF-CONSCIOUSNESS

"Nobody ever *was* or *had* a self. All that ever existed were conscious self-models that could not be recognized *as* models. . . . You don't see it. . . . This is not your fault. Evolution has made you this way."[1] There is now something of a cottage industry producing such news items and, in the process, rendering the problems of consciousness and self-consciousness even more mysterious than might have been thought. "Evolution"—the great magician—works silently and cleverly to create the real sense of self in a "system" (the nervous system, of course) that possesses no "self" as such but merely a "model" of one. What one is aware of is not what is there. Awareness features a self; reality contains only models. Were it otherwise, science and philosophy might have to contend with the dread Cartesian *ego*. But, thanks to advances in something or other—the "brain sciences," presumably—it is only a *model* that needs explaining. It is doubtful in the extreme, however, that Descartes would have been chastened by arguments along these lines for, as noted, the *Cogito* is not at base an ontological argument but an epistemological device designed to defeat skepticism. That evolution has deceived us is precisely the sort of possibility Descartes considered in relation to the evil demon. Now the deception takes the form of having one accepting one's *self* as a continuing and uniquely identifiable entity, whereas all that really exists is some sort of model of such a thing.

It is not absorption into the brain sciences that spells doom for older and common-sense notions but the penetrating light of conceptual analysis. Wittgenstein is again the figure of consequence. One of his most original disciples, Elizabeth Anscombe, grasped the significance of self-consciousness and set out to reduce it to philosophically manageable dimensions.[2] To speak of self-consciousness is to speak of one being conscious of something, and to know in some sense of "know" that it is oneself. All sorts of sensations and actions have a locale. What is it that allows Fred to ascertain without doubt that it is he, Fred, who is the source of these events? Anscombe posed this as a real question with a real answer: "with what object is my consciousness of action, posture, and movement, and are my intentions connected in such fashion that *that* object must be standing up if I have the thought that I am standing up and my thought is true? And there is an answer to that: it is this object here."[3]

The problem—the real problem of "Cartesianism" if there is one—arises not from a two-substance ontology or from introspection as a mode of inquiry. The problem is that of the "I" when rendered metaphysically distinct from spatiotemporal embodiment. If there is no strict identity between Descartes and his materiality, then, although it is true that Descartes can be deceived by all material modes of knowing, it is also the case that he can be deceived in like manner as to being "Descartes." After all, the entity identified as Descartes is a material object, too, which seems to match up with specific spatiotemporal conditions. If all Descartes were to mean by "I" in the *Cogito* were his extended body, then the deceiving demon's success would be complete. The way "I" works is different from ordinary descriptions and inferences. Imagine being filmed performing a complex task under hypnosis and while wearing a disguise. Later, now fully conscious, imagine watching the film and being aware (conscious) of all the filmed actions. The person of which the viewer is conscious is, as it happens, the viewer herself. But this is not *self-consciousness*. No amount of further information of this sort would enhance self-consciousness or even reach it. In one sense, the right answer to the question "who are you viewing?" would be "myself," but this would be a matter of discovery, not direct and immediate awareness. Thus "I" is not identical to a spatiotemporal array of elements substitutable for "I" by any name. One might be shown a film of, say, a half-dozen such disguised figures, one of them being the viewer filmed earlier and under hypnosis. If "I" were no more than a referring term, then, as Anscombe observed, "it would be a question what guaranteed that one got hold of the right self, that is, that the self a man called 'I' was always connected with *him*, or was always the man himself. Alternatively, if one said that 'the self connected with a man' meant just the one he meant by 'I' at any time, whatever self that was, it would be by a mere favour of fate that it had anything else to do with him."[4]

How, then, is "I" to be understood? It is to be understood as a primitive, a (Reidian) common-sense principle expressing itself linguistically in different ways in different cultures but functioning neither referentially nor conceptually. "What do you *mean* by 'I'?" is not a question that is answered referentially; e.g., by pointing to a picture of oneself or to one's reflection in a mirror or to one's DNA profile. It is instead a term of acknowledgment, an expression that designates not place or time but *responsibility*, even in such innocent locutions as "I like strawberries." If "I" had to *mean* something, then, as with meaning in general, its locus would be in the world of practices and conventions such that one would have to learn from that world just how to apply it. True, we all learn how to use words, but, once properly instructed, we reserve to ourselves final authority on the use of "I," and perhaps on no other word. But there is more to say about externals.

EXTERNALISM AND ANTI-INDIVIDUALISM

As noted in the previous chapter, yet another seemingly decisive blow against "Cartesianism" was delivered by Hilary Putnam in his 1975 essay "The Meaning of 'Meaning,'" which has as its major conclusion the *externalization* of meaning.[5] Putnam's essay is the *locus classicus* of that Twin Earth that philosophers have now been visiting for a quarter of a century, often approaching it as something of a shrine. For those whose travels have missed this site, a hasty tour is sufficient.

Twin Earth is a replica of Earth in all details except one. It is inhabited by persons like ourselves, speaking our language, sharing our beliefs and attitudes, dressing as we do, seemingly knowing the things we know. They might easily be our friends and neighbors. The one difference is this: What on Earth we refer to as "water" is a liquid formed from the combination of two hydrogen atoms and one atom of oxygen, H_2O. The colorless liquid that is called "water" by the occupants of Twin Earth, however, is formed by the atoms XYZ. Thus, what we "mean" by the utterance "water" cannot be what is meant when Twin Earth residents utter the same word. From this, the argument proceeds to the conclusion that meaning, as Putnam says, "ain't in the head" but is formed in the external world that shapes and fixes language and thought. So much, then, for that Cartesian *self*, that *res cogitans* so imperturbably certain of its own thoughts to which special ownership and privileged access were the defining marks. If to be a thinking thing there must be thought, and if thought is about or means something, then a thinking thing is one possessed of what is supplied from the outside, not what is found "in the head." On this understanding, Descartes was just misled by an introspective mode of inquiry that guaranteed its own results.

What Twin Earth data require us to reject is what has been referred to as a mentalistic *individualism*, described by Tyler Burge this way: "In the elderly Cartesian tradition, the spotlight is on what exists or transpires 'in' the individual—his secret cogitations, his innate cognitive structure, his private perceptions and introspections, his grasping of ideas, concepts or forms."[6] Against this "elderly" tradition, Burge arrays the insights of Wittgenstein and others who have drawn attention to the ineliminably social and discursive sources of meaning such that the individual, to be thinking of anything, must be employing resources that originate in social space. There is no genetically predetermined code or interior representation for a Jeep Grand Cherokee, so if Tom is thinking about the one now being serviced he is thinking about an entity supplied from the outside. Before the advent of automobiles no such thought, or at least not that precise thought, was possible. In a word, thinking is contextual and the content of thought is culturally *given*, not introspectively *found*. The Cartesian just missed the boat here, thinking it was located in that private duchy directly accessible solely by the individual *res cogitans*. Actually, it was at the public docks all along.

However, when we turn to part 2 of Descartes' *Discourse on Method*, we read these lines:

> I had become aware, even so early as during my college life, that no opinion, however absurd and incredible, can be imagined, which has not been maintained by some one of the philosophers; and afterward in the course of my travels I remarked that all those whose opinions are decidedly repugnant to ours are not on that account barbarians and savages, but on the contrary that many of these nations make an equally good, if not a better, use of their Reason than we do. I took into account also the very different character which a person brought up from infancy in France or Germany exhibits, from that which, with the same mind originally, this individual would have possessed had he lived always among the Chinese or with savages. . . . I was thus led to infer that the ground of our opinions is far more custom and example than any certain knowledge.[7]

Recall what the *Cogito* was all about in the first instance: it was not the basis on which we correctly identify water or woodchucks. A person brought up from infancy in France will later say "*rouge*," whereas the native-born German says "*rot*" and the Londoner says "red." The dichromat might well say "gray." Obviously, "gray" cannot *mean* the same as "*rouge*" or "red." But the percipients in all these instances have indubitable conscious awareness, no matter how confused or misled or "contextualized" their knowledge claims prove to be.

To the extent that there is any sort of "Cartesian theater of the mind" to be found in Descartes' philosophical works, nearly everything featured originates in the world beyond the mind, not in it. Thus, it is not the content of the performance that matters but the radically different position of the two audiences. One can know of the performance only through descriptions passed on by that one person who is in the theater and not listening at the door. In a word, *it's not about meaning*, and therefore it's not about H_2O or XYZ. The relevant consideration is not the provenance of meanings but the certainty attached to the fact that one is thinking of anything.

Perhaps an illumination of this otherwise subtle point is provided by investigations of unilateral dichromatism, a rare condition in which the percipient has normal color vision in one eye but with significant loss of color sensitivity in the other.[8] The percipient in this case might be seen as the simultaneous occupant of Earth and Twin Earth, depending on which of her eyes is open to the world. With her trichromatic vision, she reports what is customarily reported by others exposed to precisely controlled wavelengths at photopic levels of illumination. With her dichromatic vision, she reports all wavelengths above 502 nanometers as "yellow." It is unarguable that her use of words such as "yellow" and "gray" is culturally sanctioned, but it is comparably unarguable that the radically different visual *qualia* are inexplicable in cultural or discursive terms. This same research has applications to Frank Jackson's famous "Mary problem."[9] Mary knows all there is to know about the physics of light, about optics, visual systems, pigment chemistry, etc.

However, having been reared in a black-and-white room, she has not had the experiences arising from visual stimulation in the real world of colors. She now enters that world and, to put the point directly, *learns something*. Had all of physics and physiology been sufficient, her entry into the world of color would have been redundant. As Jackson states the case: "If physicalism is true, she knows all there is to know. For to suppose otherwise is to suppose that there is more to know than every physical fact, and that is what physicalism denies."[10]

It might be argued that the phrase "knows everything about X" carries with it *everything*, and presumably this would have to include the consequences of entering the illuminated world. But there is a difference between knowing by acquaintance and knowing by description (as Russell was at pains to insist),[11] and what Mary might have learned *about* experience would not match up with what she learned *from* experience. The subject of the research by Graham and Hsia, had she gone through life with the color-sensitive eye patched, would have had a "Mary" moment once that patch was removed, though she had seen the world in its variety (but without color) for years.

With this in mind, we begin to see that there is nothing whatever in the "discovery" that meaning is socially or culturally constructed (and who ever doubted it?) that bears on the adequacy of the *Cogito* or the conclusions Descartes drew from it. The residents of Twin Earth believe that water is XYZ, those of Tucson and Bicester that it is H_2O. In relation to Descartes' maxim, it would matter not one whit were the phrase changed from *Cogito ergo sum* to *Credo ergo sum*. The reality of my "being" is as assured by the act of belief as by that of thought, belief here being one species of thought. Opinions and beliefs are shaped by culture and custom, by local conventions and by the provincial narrowing of the imagination. Janet on Earth, when thinking of water, is, in fact, thinking of H_2O, whereas Sam on Twin Earth is thinking of XYZ. But both are thinking, and that much is certainly "in the head," for it is surely not on the periodic table. There is even a bit more to it than this. The credit given to Twin Earth's vindication of externalism may well be in an overvalued currency, depending on just what sort of philosophical tool is needed for the chosen problem. If Janet and Sam are the participants, we must be clear on the difference between (1) what each of them is *thinking* when the physical object associated with thought is H_2O for Janet and XYZ for Sam, and (2) what each of them *means* when talking *about* or, for that matter, thinking *about* the object. What are we prepared to conclude from what Janet and Sam "mean" when speaking of "water" or of water? In point of fact, just what is the burden of the question, "What does it mean to say that my word 'water' *means* something?"

On Twin Earth, the real chemical compound is not water but *t*water. On Earth, it is water. If Putnam is right, just in case Janet and Sam both claim to be thinking of water, one of them is not; one of them is actually thinking of what in fact is *t*water and so the same word cannot *mean* the same thing if it refers to a physical object. Suppose, however, the term is understood as referring to the phenomenological *effects* reliably associated with the object. Assume that for both Janet and Sam the word "water" just *means* a transparent, potable, odorless liquid delivered through pipes and taps and providing refreshment in hot weather. In that case, H_2O and XYZ do have the same meaning, for both map on to the same phenomenological space. What Janet and Sam consciously apprehend is a complex sensation chiefly characterized by the properties *wet*, *transparent*, *odorless*. On Earth, we call it water. On Twin Earth they call it water. Ah, but on Twin Earth that which is *wet*, *transparent*, *odorless* is not H_2O. Fine. Shall we use another name? But by any name, it remains *wet*, *transparent*, *odorless*, and it is this assortment of properties that Sam and Janet would hope to convey with a word. To the extent that a word stands as a sign for a fixed physicochemical entity, water and *t*water have different meanings. To the extent that a word stands as a sign for effects readily predictable from the phenomenological properties of the entity, water and *t*water have the same meaning.

The confusion here seems to arise from the assumption that references to two distinct entities cannot have the same meaning. At the level of empirical fact, this is arguably the case, though not across the board. It will depend on how "distinct" and "entity" are to be understood. Max thinks that Phosphorus is bright and Phil thinks that Hesperus is dim. If the "entity" in question is the visual experience had by each, then they are not referring to the "same entity." If it is instead Venus that is referred to, then "it" means different things to Max and Phil, though it is the same entity. These are complications on which the Sophists of old earned a good living, but they really don't do the sort of work required of a convincing externalist account. Suppose the subject in the Graham-Hsia study views Venus with one eye, then with the other eye, giving separate accounts. Here each is an account of the same planet and a different account of the experience mediated by the separate eyes. It is unclear where in these accounts "meaning" enters or even where the Fregean distinction between sense and reference arises, for *qualia* really don't *mean* anything. Their ontological status is not gained by way of meaning and therefore not by way of acculturation or the "linguistic turns" thereby set in motion.

Before leaving Twin Earth, at least a few words are in order concerning the metaphysical plausibility of the place. A detailed analysis falls beyond present purposes but, being a realist about "water," I am obliged to deny the residents of Twin Earth the right to use "water" for their pools of XYZ. Nor is their Earth actually the "twin" of ours. If XYZ is not H_2O, then at best we have siblings, not twins. Note that fraternal twins are no more alike genetically than any pair of siblings of the same gender. Where "water" is something different from *water*, we can be sure there are many other and surely profound differences. That the same word may be used in widely different senses—from the "cheesy" taste of cheddar to the "cheesy" design of modern appliances—does not bear on the question of the properties by virtue of which something is CHEESE; e.g., derived from milk, cultured by microorganisms, etc. Before the age of modern chemistry, no one on our own Earth "meant" H_2O when referring to water nor, to this day, do very young children or uneducated adults. It would be unnecessarily adventurous, however, to conclude from this that "water" has some shifting meaning *at the phenomenological level* from age to age, epoch to epoch. It would be rare, especially these days, to have the pleasure of pure H_2O, though we all shower, boil eggs, wash the car, feed the flowers, and fill the birdbath with—water. The point, of course, is that *externalism* is not the last word on the matter of meaning, *pace* Wittgenstein, Putnam, Rorty, et al. Meaning is very much "in the head" once we accept that it is what Sam *means to convey about his experience* rather than whether he found the right word for the physicochemical object with which he is in conscious contact.

However, it is not just this aspect of externalism at issue here. Instead, it is the inference or conclusion that, owing to the "external" fixing of meanings and to the relationship between consciousness and thought, "consciousness" itself is externally fixed and fashioned, this then being a short step away from *the social construction of the self* and the elimination of all sorts of mental entities from "the head." It is a variant of externalism that replaces the Cartesian "self" of the *Cogito* with a socially constructed "self" arising as a cultural artifact whose psychological properties are discursively bounded. Farewell to *Ego*! I will argue here that the inference would be unwarranted even if externalism succeeded, and is highly misleading in any case. If we regard consciousness as a phenomenon directly accessible solely by introspection and thus internal, we can (for the moment) delete the introspective part and analogize it to, for example, digestion. If we then take meaning to be a property created by social practices, we might carry the analogy forward by treating it as, let us say, candy. Digestion refers to the action of chemicals on ingested food. Consciousness refers to the person's "action," as it were, on the contents of that person's unique experience. If all experience were restricted to events external to the body, it would not follow that consciousness itself is "an artifact of culture" any more than digestion is "an artifact of candy." The analogy crudely highlights what is defective in those arguments that reduce consciousness to a social construct. Why do serious and informed philosophers defend the view? Chiefly to avoid that "Cartesian" *self* seen as Ryle's *ghost in the machine*. As with Prospero to Ferdinand, modern philosophy would have us cheerful in the news that our very *selves* —those *essential 'personal identities'* uniquely possessed and with their own inner mental lives "were all spirits and are melted into air." But were they? Have they?

PERSONAL IDENTITY AND THE SELF

A convenient entry here is by way of the Ship of Theseus.[12] In his "Life of Theseus," Plutarch cites unnamed ancient philosophers vexed by an aspect of the Theseus myth. Considering the old planks replaced by new wood on the ship Theseus had sailed to Crete, they wondered at what point the refurbished ship was no longer "the ship of Theseus." The problem, as modern as it is ancient, is that of the continuity of identity for anything that undergoes change; the problem of when the "essence" of a thing no longer obtains, such that the thing is no longer the sort of thing it was.[13] How can it be said, for example, that between conception and death at an advanced age, the entity in question retains a continuing identity? For to answer that nothing else ever occupied the space this entity has occupied at any and every instant of time would merely establish a continuity of mass or volume and surely not a continuity of identity. Among the several legacies of so-called Aristotelianism that the modern age of science took to be counterfeit, the very concept of *essence—ousia—* was subjected to sustained criticism. This is a long and interesting chapter in the history of ideas but one beyond the purposes of this book. It is sufficient to pick up the account with Hobbes and then with Locke's well-known distinction between what he called the *real essence* of a thing and its *nominal essence* and then briefly consider the distinction in the context of personal identity.

Hobbes's *De Corpore* was published in 1642, the English version appearing in 1675 as *De Corpore Politico*. In the chapter "Of Identity and Diversity," Hobbes moves directly to the Ship of Theseus:

> For if . . . that ship of Theseus, concerning the difference whereof made by continual reparation in taking out the old planks and putting in the new, the sophisters of Athens were wont to dispute, were, after all the planks were changed, the same numerical ship it were at the beginning; and if some man had kept the old planks as they were taken out, and by afterwards putting them together in the same order, had again made a ship of them, this without doubt, had also been the same numerical ship with that which was in the beginning; and so there would have been two ships numerically the same, which is absurd.[14]

Two distinguishable entities cannot be the same. Hobbes must argue for a sense of identity that does not collapse into absurdities. There is, of course, a sense in which the water in the river is the same as the water now forming the cloud. There is a sense in which it is different. There is a sense in which Socrates as child is different from Socrates as man. Regarding personal identity, what is the sense in which it has continuity amid the corporeal fluxes of the hour and the lifetime?

> That man will always be the same, whose actions and thoughts all proceed from the same beginning of motion, namely that which was in his generation; and that will be the same river which flows from one and the same fountain, whether the same water, or other water, or something else than water, flow from thence; and that the same city, whose acts proceed continually from the same institution, whether the men be the same or no.[15]

Here the ever mechanistic thinker accepts a continuity of the person throughout the seasons of bodily change so long as one condition is satisfied: *the same beginning of motion, namely that which was in his generation.* Rejected are Aristotelian "essences" in favor of the identity of functional states. All in all, Hobbes's treatment would set the stage for Locke's analysis of the issue and his own conclusions regarding the continuity of the material *man* but the alterations of that man's *person.* To see how Locke attempts to the render all this conformable to the science of his time, we begin with his antiessentialism.

In *An Essay Concerning Human Understanding,*[16] Locke develops the distinction between "real" and "nominal" essences. The central thesis is developed in book 3, chapter 6.[17] Here he notes that any number of creatures of his general form, or of considerably different form, might have more or better faculties than he, but that such differences are not in any way "*essential to the one or the other, or to any individual whatever, till the mind refers it to some sort or species of things.*" What will give Locke or anyone else that continuing identity that might have been incorrectly regarded as one's "real essence" is no more than that on which the various habits and dispositions of the mind settle.

Following Hobbes, Locke gave the problem of personal identity its modern formulation. One of the aims of *An Essay Concerning Human Understanding* was precisely to challenge *essentialism.* Faithful to Newtonian science, Locke regarded the *real* essence of a thing to be beyond the powers of sense, grounded in its fundamental corpuscular structure, and manifesting itself, if at all, only at that perceptible level on which variable and context-dependent *nominal essences* are fashioned. Locke's *real* essence is a congeries of submicroscopic particles held together by gravitational forces but perceived in ways that generate such *nominal* characterizations as "physician friend to the Earl of Shaftesbury," "Fellow of the Royal Society," "a rational animal." These characterizations arise from conventional discourse, the contingencies of culture and context, and the nuances of perception, memory, and mental life.

It would be fair to say, in light of Locke's reference to the *real essence* of things, that he is an ontological essentialist but agnostic at the level of epistemology. Even if science were to attain perfected methods of measurement with which to identify the *real* essence of an item or entity, it would not match up with the manner in which we come to know and describe it.

Consistent with the distinction between real and nominal essences, Locke reduces personal identity itself to the merely contingent contents of consciousness such that the notion of an *essential* self gives way to an identity that is hostage to the limitations of memory and the vagaries of experience. There is an intimation of the *real* versus *nominal* distinction in Locke's famous example of the Prince and the Cobbler: As each sleeps during the night, the contents of their consciousness are switched. On awakening, Locke grants that each will be "the same man" but *not the same person*. Each remains the same man, the reference here to a physical body. But the *person*—as that person would disclose what he takes to be his "essential self"—is but what a series of recollected experiences constitute as being a prince or a cobbler. The real essence of neither is disclosed by such nominal features as a princely bearing or special skill in turning leather. Locke is clear on this point. Personal identity consists, he says, "*not in the identity of substance, but, as I have said, in the identity of consciousness, wherein if Socrates and the present mayor of Queinborough agree, they are the same person: if the same Socrates waking and sleeping do not partake of the same consciousness, Socrates waking and sleeping is not the same person.*"[18]

Contemporary readers might be less challenged by such notions than were Locke's contemporaries, for the issue that many judged to be at the very bottom of the matter was the eschatology of the Christian religion. Just who or what "survives" the death of the body, and in what form, were understood to be not merely academic questions. Some of Locke's friends and colleagues had already cast the issue in scientific terms and thus further troubled the defenders of orthodoxy. Robert Boyle, for example, had applied corpuscularian physics to the claims of scripture and found the latter to be wanting. He presents the main argument with uncompromising efficiency:

> When a man is once really dead, divers of the parts of his body will, according to the course of nature, resolve themselves into multitudes of steams that wander to and fro in the air; and the remaining parts, that are either liquid or soft, undergo so great a corruption and change, that it is not possible so many scattered parts should be again brought together, and reunited after the same manner, wherein they existed in a human body whilst it was yet alive. And much more impossible it is to effect this reunion, if the body have been, as it often happens, devoured by wild beasts or fishes; since in this case, though the scattered parts of the cadaver might be recovered as particles of matter, yet already having passed into the substance of other animals, they are quite transmuted, as being informed by the new form of the beast or fish that devoured them and of which they now make a substantial part.[19]

Perhaps Locke's reduction of personal identity to the contents of consciousness was an attempt to preserve the Christian canon by permitting "persons" to be reconstituted in any material form sufficient for there to be consciousness. The celebrated exchange of letters (1695–1697) with Edward Stillingfleet, Bishop of Worcester, leaves no doubt about the importance churchmen attached to this part of Locke's theory.[20]

But if the Locke of the *Essay* of 1690 is "metaphysical" on matters of eschatology, the author of the *Two Treatises of Civil Government* seems more comfortable with what is "essentially" human. The *Treatises* were among the most widely reprinted works of the eighteenth century. In them, the author looks less to Newton than to Stoic philosophy, arguing that "*the Law of Nature stands as an Eternal Rule to all Men,* Legislators *as well as others. The* Rules *that they make for other Mens Actions, be conformable to the Law of Nature, i.e. to the Will of God, of which that is a Declaration, and the* fundamental Law of Nature *being* the preservation of Mankind, *no Humane Sanction can be good, or valid against it.*"[21] To invoke a universal law of nature as dispositive in human affairs is to extend Newtonian "covering laws" into regions vastly more cluttered and dynamic than celestial mechanics. Moreover, it is to assume something about human nature itself that goes beyond social conventions and cultural nuances. It is, indeed, to assume something *essential* about human nature, against which human sanctions themselves cannot safely or validly prevail.

ESSENTIALISM

What I hope to draw out from these few lessons from Locke is a measure of the costs incurred when essentialism is either abandoned or is reduced to a merely nominal essence supporting the more pernicious forms of moral and juridical relativism. The seasonal hibernation of essentialism is based on cogent criticisms warranting at least a judicious summary and, I hope, judicious appraisal. Its reappearance, however, must be routine if there is to be the very taxonomy of problems for the special sciences to address.

Let us begin with the traditional sense of something being a certain "kind" of thing in that it has *essential* properties. To speak of something in these terms, at least as Aristotle would intend this to be understood, is to speak of what it is to be a certain kind of thing. As Aristotle used the term, "essence" is conveyed by the phrase το τι ειν ειναι—the "what it is to be" something. But the "something" falls under a universal category. Thus, a cat is an animal, and were there no such *taxa* as plants and animals, there could be no cats. This is not to be confused with Platonic notions of "true forms" thought to have ontological standing apart from their actual instantiations. Rather, Aristotle—the great natural scientist and relentless taxonomist—recognizes degrees of kinship as expressed in the form of similar appearance, similar modes of behavior, similar responses to the environment. Thus, the sense in which "Coriscus is a man" is different from the sense in which "Coriscus is musical." Coriscus is *essentially* an instance of "what it is to be" a human being, but merely accidentally musical, along the way. As for the το τι ειν ειναι of human beings, it is finally to be a rational animal, disposed naturally to social and political modes of life and able to base actions and choices on rational deliberation. The το τι ειν ειναι for this kind of creature includes centrally *fitness for the rule of law*. It is to be a specific "kind" of creature, arising from the natural order of things, as does the *polis* itself. To be "essentially" something is to be so by nature, as in the pathetic example of the δουλος φυσικοσ, the "slave by nature." It is to be a *natural kind* of thing.

We begin to see the stress points in such accounts when we ask just how stable are those general classes that are taken to be "natural kinds." Alexander Bird gives a hint as to the complexity of the problem:

> When one visits a greengrocer, in the section devoted to fruit one will find, among other things, apples, strawberries, blackcurrants, rhubarb, and plums, while the vegetable display will present one with potatoes, cabbages, carrots, tomatoes, peppers and peas. If one were to ask a botanist to classify these items we will find rhubarb removed from the list of fruit and tomatoes and peppers added. . . . Following this line . . . one might conclude that there really is no absolute sense in which there is a natural classification of things into kinds.[22]

A way out of this bind is to seek a more modest principle of classification, such as Locke's notion of *nominal essence*. The term "apple" ranges over a variety of items with a set of attributes understood to be required if the term is to be applied correctly. An entity has the *nominal essence* of "apple" just in case it possesses the shape, color, taste, size, etc. commonly found in things growing on trees of a certain kind and not on other kinds of trees. It should not seem odd to identify one sort or a variety of items with a set of attributes understood to be required if the term is to be applied correctly. That is to say, it should not seem odd that designating one sort of entity as a "natural kind" is often parasitic on still other stipulations of the same sort. Thus "apples," among other considerations, grow on, yes, *apple trees*.

Are we not still in the bind? The problem here, of course, is that nominal essences are controversial in just the way that "natural kinds" are, when dependent entirely on such descriptions. Saul Kripke's "tiger critique" illustrates this: To say that the tiger has a nominal essence that includes "large, quadrupedal, carnivorous, black and yellow cat populous in India" and that this qualifies tigers as a natural kind is to fall prey to what Kripke calls *epistemic credulity*. Those who first saw and reported on such creatures could have suffered from visual defects, or may have seen only the few tigers who actually do eat meat, or might have failed to see a fifth leg on half the specimens.[23] Descriptions, no matter how consistent, are fallible accounts of *what is there* and therefore cannot be the last word on even on the nominal, let alone the "real," essence of anything.

In this connection, the formidable critiques of scientific realism such as those mounted by Nancy Cartwright and Bastian van Fraassen cast doubt on any confident assertion that the laws of science have real standing apart from scientific modes of representation and cognition. Cartwright's "simulacrum theory" reduces scientific explanations to what is no more than a model that permits the derivation of what she calls "*analogies* for the messy and complicated phenomenological laws which are true of it."[24] On accounts of this sort, notions of "essences" or of "natural kinds" can only be theoretical or pretheoretical models, useful as a way of analogizing from one level of observation to another, the choice of levels based rather on pragmatic considerations than on any privileged means by which to comprehend ultimate reality.

Against this line of criticism there are persistent reminders that reality and our cognitive access to it are not finally reducible to culture and context, and almost certainly not confined to the realm of the observable. That there has never been found a mile-high mound of gold is nothing but a fact of experience, as corrigible as empirical descriptions of what it is to be a tiger. That there has never been found a mile-high mound of uranium is known a priori, given the laws that limit the cohesive bonds composing this element.

That the gravitation laws find confirmation in observations and measurements is a finding that reflects at one level that model of explanation and mode of comprehension characteristic of the "scientific way." This, however, would scarcely warrant the conclusion that the gravitation laws are but grammatical or cultural constructs. Nor is it telling against essentialism that it arises from a set of cognitive dispositions widely shared across time and cultures. If, indeed, there is a nearly ubiquitous disposition not only to describe or understand nature in essentialist terms, and if this disposition has proven to be a necessary feature of adjusting to the physical affordances of nature, then the burden would shift to the antiessentialist.

Here one faces core issues in philosophy of science and philosophy of physics well beyond the scope of this chapter. It is sufficient to note, however, the gradual movement of thought back in the direction of "Aristotelian" conceptions of natural powers and properties inhering in things and thus accounting for their activity. At the end of the day, the pragmatic dimension of cognition is decisive, though arguments will abound as to how close we are to the end of the day and as to what the measure of efficacy is to be.

If metaphysical defenses of essentialism are strained, the same is not the case with what might be called psychological or ethological defenses. Whether or not there are real essences and fixed natural kinds recent research leaves no doubt but that children and adults habitually perceive and organize aspects of the external world in essentialist fashion. The tendency to "essentialize" seems to be a very general cognitive function. There is a substantial literature addressed to "psychological essentialism" based largely on evolutionary principles establishing the adaptive advantage gained by this mode of cognition. The processes at work are not of the descriptive empirical sort. What one regards as the "essence" of a thing is not exhausted by its observable features alone but arises from a conception of some internal, intrinsic feature that causally brings about these features.

A recent review of this literature has been provided by H. Clark Barrett.[25] He identifies the principal features of essentialism as these: First, there is the concept of "executive causation," the concept of a central cause that accounts for the observed properties of an object. Ruled out by this concept are properties that are merely correlated with the object. One might say that in taking the musicality of Coriscus to be an accidental feature but his humanity as an essential one, Aristotle exemplifies the "executive causation" concept. Next, psychological essentialism requires the feature of what has been called "rich inductive potential."[26] Having established the "essence" of the thing, the observer is able to anticipate any number of actions and features not now or yet observed but readily inducible from, alas, the "essence" of the thing. As Clark Barrett says, through the rich inductive potential conveyed by essentialism, the observer reckons that "there is more to the nature of a kind than that which we use to recognize it . . . the executive cause has a potentially limitless number of regular effects."[27]

Yet another adaptive feature of psychological essentialism is that it disposes the percipient to classify objects according to potential effects even while remaining ignorant of that "executive cause" by which the effects are brought about. The example offered by Barrett is that of food types that are poisonous (e.g., mushrooms), animal kinds that are "predatory," etc. The adaptive advantage gained by this mode of cognition is obvious.

The "essentializing" disposition is not some relic of an age of superstition and ignorance, but rather a pervasive heuristic on which success in the routine affairs of life depends. Where the arguments of the metaphysicians come down more or less equally on the various sides of a question, the pragmatist need not be so humble as to withhold his own validating criteria. And, to the extent that data of any sort might be admissible even within the often antiseptic arena of metaphysical speculation, it would seem that the record of human social and political organization might be adduced in support of the proposition that, exceptions duly noted and explained, "man" seems very much and essentially to be a rational creature disposed to live in the company of others and to ground the patterns of social interaction in principles of a recognizably moral nature. Unless philosophically tutored, such creatures are not plagued by "the problem of other minds" or the social construction of themselves for, given the ubiquitous cognitive disposition to "essentialize," they recognize in the actions and general form of other human beings the operation of some sort of "executive cause" that must be like the one by which their own actions proceed. And, owing to the "rich inductive potential" borne by an irreducibly essentialist conception of human nature, they are able to predict a wide range of actions under the virtually limitless conditions that might at some point be faced by themselves and those like themselves. In these respects, if not in metaphysically more ambitious ones, it would seem safe to say that persons are essentially *themselves*, not as empty tautologies but as self-aware (conscious!) beings. This would seem to be the necessary condition if social forces are to construct *anything* of psychological moment.

There is, however, more to it than this, for persons are not simply and essentially themselves; they understand themselves to be essentially a certain kind of entity. This is precisely what Descartes' distinction between *res cogitans* and *res extensa* is all about. Consider again the widespread "essentializing" heuristic so necessary in the business of daily life. At the center of that business are associations and confrontations with others relevantly "like" oneself in outward appearance. Now consider the nature of such interactions just in case one were to reserve to oneself alone the essential property of consciousness or awareness or, more generally, mental life. But consider also the perils courted by one who attributes such a property willy-nilly. It is by the pragmatic calculus that the aptness of essentializing is measured. The last word on the validity of practices and habits is not sounded by the metaphysician but by the facts of the world as these practices are brought to bear on them.

PERSONAL IDENTITY (CONTINUED)

Having asserted above that persons are not simply and essentially themselves, I am of course drawn into a discussion of "persons," for if persons as such are fictive then they are not *essentially* anything. Locke, as we have seen, rejected substance theories of personal identity in favor of the persistence of certain ideas in consciousness. What makes the cobbler a cobbler is different from what makes him a man. What makes him *that person* are those experiences accessible to memory and conscious reflection such that were the contents transported to a different body (e.g., that of a prince), the new body would be the personification of the cobbler.

With Locke, there is still a narrator of sorts, a "person" reflecting on certain mental contents. A far more radical thesis is advanced by Hume. In *A Treatise of Human Nature*, Hume effectively eliminates the person as an ontological item, replacing it with bundled perceptions held together by the laws of association. He centers his criticism of the traditional account on the shifting nature of all sensations, changes so dynamic and incessant as to be incapable of preserving a continuing personal identity. He writes: "There is no impression constant and invariable. Pain and pleasure, grief and joy, passions and sensations succeed each other, and never all exist at the same time. It cannot therefore be from any of these impressions, or from any other, that the idea of self is derived; and consequently there is no such idea."[28] Leaving no doubt as to the implication to be drawn from the criticism, Hume then famously adds:

> For my part, when I enter most intimately into what I call myself, I always tumble on some particular perception or other, of heat or cold, light or shade, love or hatred, pain or pleasure. I never can catch myself at any time without a perception,

> and never can observe any thing but the perception. When my perceptions are removed for any time, as by sound sleep, so long am I insensible of myself, and may truly be said not to exist.[29]

"Persons" on Hume's account can be no more than "a bundle or collection of different perceptions, which succeed each other with an inconceivable rapidity, and are in a perpetual flux and movement."[30]

There is no "person" within that inner theater, observing experiences on the mind's screen. There are but the perceptions themselves. As in a parade formation, where one marcher might leave only to be replaced by another, so, too, the chain of associated perceptions remains connected even as first one and then another link is added or removed. Hume encourages the reader to think of the simplest of animals—an oyster, in his example—having but one perception. The mind of such a creature would be exhaustively described by reference to that single perception. Adding other perceptions does nothing to alter the basic thesis that "personal identity" is exhaustively covered in terms of bundles of sensations.

There is more to all this, for Hume was aware that the account was more successful as criticism than as explanation. There is a connectedness to the experiences one has, this giving rise to the feeling or belief that a "person" is the necessary entity if these experiences are to inhere. But Hume's account of *causal* connections (in)famously rules out knowledge of causal powers. Strictly empiricist, Hume insists that we are able to perceive directly the temporal sequences of events but not any "cause" that explains or accounts for them. One billiard ball collides with another and the second moves. This is all apparent. What is nowhere to be seen is the alleged "cause" of the motion. Causal concepts arise from the "constant conjunction" of events of type A and events of type B. It is a natural tendency of percipients under such circumstances to come to regard such events as causal, though there is no direct empirical evidence of causation itself. For Hume, then, "personal identity" is but the cohesion of chains of perceptions. He is admittedly at a loss to explain how all this takes place, but is satisfied that there can be no more to "self" than this very process.

A latter-day Humean account of persons has been developed by Derek Parfit in part 3 of his *Reasons and Persons.*[31] According to Parfit, "what matters" is just the continuity of mental states and processes, not some arguable, perduring "person." Even granting in some sense that there are persons, Parfit argues that it is surely possible to develop a complete description of reality without invoking the concept or the ontological reality of persons. He then attempts to "deflate" personal identity by presenting various thought experiments, including the removal of both halves of the brain and locating each half into a new body, concluding that the most plausible expectation is that the original donor is no more, and that neither recipient can be the same "person" as the donor. Needless to say, the bearing such experiments have on the issue of personal identity depends entirely on the adequacy of neurological accounts of personhood. As the mind-body problem is still (to say the least) problematic, it is probably prudent not to assume its solution in attempting to settle another vexing and quite different problem. David Wiggins has drawn attention to still other difficulties with Parfit's position. To speak of recipient bodies as "remembering" having done X or Y in the past is problematic. As Wiggins notes, there is a real connection between remembering Xing and, indeed, having Xed. As the recipient bodies in the thought experiments have done no such thing, it is unclear that the notion of a transfer of memory is valid.[32]

Parfit attempts to defend a revised version of Locke's memory theory of personal identity by adjusting it to neutralize one obvious criticism. On Locke's view, just in case Person 2, at a later time, has the experience-memories held at an earlier time by Person 1, then Person 1 and Person 2 are the same person; they have the same personal identity. Clearly this will not do needed work, for it is obvious that someone remains the same person even when any number of specific experience-memories are lost. Parfit would repair the theory thus:

> I shall appeal to the concept of an overlapping chain of experience-memories. Let us say that, between X today and Y twenty years ago, there are *direct memory connections* if X can now remember having some of the experiences that Y had twenty years ago. On Locke's view, this makes X and Y one and the same person. Even if there are no such direct memory connections, there may be *continuity of memory* between X now and Y twenty years ago. This would be so if between X now and Y at that time there has been an overlapping chain of direct memories.[33]

The failure of this gambit can be illustrated by a modification of Parfit's own "teletransportation" thought experiment. Instead of a Martian device into which Parfit's desires, knowledge, beliefs, and memory have been transported, let us propose an even more clever device—call it a PARFITON—tethered to Parfit, equipped with the means by which to experience and remember in whatever way Parfit does, but with a ten-minute lag between Parfit having the experience and the PARFITON having it. Furthermore, it is a condition of the relationship between the two that the PARFITON cannot experience anything except by way of Parfit's experiences. Clearly, there is a qualitative and a quantitative violation of "identity" here, even though the experience memories are identical. That Parfit and the PARFITON are not numerically identical is obvious. They are also not qualitatively identical, for each of Parfit's overlapping chains of direct memory is *Parfit's own*, whereas the PARFITON's are parasitic, ownership here referring to a possession that can, as it were, be given away.

Parfit attempts to answer this objection by invoking the notion of "quasi-memories." If, as Bishop Butler noted in his critique of Locke's theory, it is part of the very concept of a memory that it refers to the experiences of the one whose memory it is, then memory as such presupposes personal identity and cannot be used as evidence for it. It is in response to this that Parfit suggests something called "quasi-memory," which he would have worked this way: Smith remembers having had an experience. In fact, *someone* did have this experience. Smith's memory of the experience is causally dependent, "in the right kind of way," on just that past experience. Now, if Smith's memory is in the form of memory traces stored by processes and regions of the brain, it is conceivable that the brain sciences at some future time would allow another to have the experience and thus to have a *quasi-memory* of what Smith had remembered. [34] But in this case the quasi-memory is not causally dependent on any experience. Rather, it is dependent on some alleged *trace* of an experience and is thus not dependent "in the right kind of way." Besides, no one remembers a "memory trace," nor has the notion ever risen higher than a vague theoretical term designed to capture the stability and then the often evanescent nature of memories. Experience leaves a (metaphorical) "trace," and, with time, the "trace" (metaphorically) fades.

We must inquire further into Parfit's particularly inventive suggestions or theories that would deny special ontological standing to "persons." Parfit on "persons" is a reductionist, accepting as actual no more than mental and physical states and the relations among them. On this account, there are only "a brain and body, the doing of certain deeds, the thinking of certain thoughts."[35] Thus, the criterion for personal identity is psychological: personal identity obtains insofar as there is psychological continuity and the abiding connectedness among recalled experiences. It is the connectedness that "matters" for Parfit, not the "personal identity" of the philosophers. As it is the connectedness that matters, Parfit is able to extend the implications of the thesis to the moral realm. The "psychological continuity" obtaining in memory and thought is a shifting commodity and may not preserve the record of good and evil in the manner of a faithful text. As he says in the matter of responsibility, "we can defensibly claim that psychological continuity carries with it desert for past crimes."[36] So, just in case this "continuity" undergoes significant alterations in the ripeness of time, the responsibility for earlier conduct is diminished in the present circumstance. There is now some sort of degree-responsibility that becomes less and less as the psychological continuities are fractured and lost. The old man no longer is as fully bound by promises made in his youth, for the intervening years have disrupted that psychological continuity on which personal responsibility depends. There are successive "selves" arising from new patterns of continuity, each next "self" less burdened by such guilt or responsibility as might have attached to an earlier version in which the means-ends chains still obtained.[37]

We see that on Parfit's account, what is proposed is a succession of "selves" but not of enduring "persons." It is Parfit's view that there is no metaphysical advantage enjoyed by a fixed "self" over and against that less than fully stable victim or beneficiary of the alterations to everyone produced in the passage of time. Rejected is the idea that "selves" enjoy some sort of metaphysical advantage not available to socially and physically identifiable entities. It is an important thesis or theory for, if successful, it would eliminate the need for conferring special authority on "self" reports and related aspects of that "inner self" traditionally regarded as distinct from the public (socioempirical) self.

It is less than clear how one goes about establishing metaphysical "advantages" and then grading them. Parfit's claim that a "self" adds no such advantage receives no support beyond some sort of ontological economy. There is more to metaphysics than ontology, however; there is also epistemology. Might some kind of epistemic advantage be gained by an entity with an "inner self" over what is available to a public socioempirical self? The answer is yes. There is Betty, seen holding her cheek and displaying a grimace. The public "self" can be characterized in various ways, each subject to error and correction. It may be thought that Betty has a toothache, or that she has bit the inside of her cheek, or that she has a burning sensation. When queried,

Betty says that she is rehearsing the part of actress in pain. Except under the most unusual circumstances, this brings the question to a successful conclusion, as would be the case had she said that she was suffering from a toothache. The epistemic authority conferred on the "inner self" is not available to the public empirical self and thus the inner self is at a metaphysical advantage.

Beyond this, is it really the case that "persons" are evanescent entities not possessing that stability and identity presupposed by moral theories and by adjudication itself? Christine Korsgaard has recognized the practical consequences of Parfit's theory and finds in them sufficient reason to be less than accommodating.[38] There are, after all, compelling reasons for retaining the reality of "persons" as entities with identities that endure over time and over the many influences bearing down on them in every phase of life. A theory that dissolves contracts and agreements in the mists of time and assigns degrees of responsibility to persons on the basis of their powers and recollections at a given time encourages the more dismissive estimations of philosophy as worthy of attention by those dealing with real issues. Thus does Parfit find himself in the same position that Locke was in when the "Scriblerians" stretched his thesis to the end of its conceptual tether and showed its implications to be preposterous.[39]

The difficulties faced by philosophers seeking to demystify personal identity or, better, that immediate apprehension of one's "self"—to demystify *self-consciousness*—extend back to Descartes' *Cogito*, to Locke's efforts, to Hume's more refined version, and to Thomas Reid's critical assessment of the entire project. It was Reid who drew attention directly to the fact that "the evidence we have of our own identity, as far back as we remember, is totally of a different kind from the evidence we have of the identity of other persons, or of objects of sense."[40] Less important than the adequacy of Locke's "memory" theory of personal identity is the profound difference at the evidentiary level between how we establish the identity of others or of perceptual objects and how we retain the awareness of self.

Of greater significance is the recognition of what would be lost just in case there were no continuing personal identity. It is too facile at this point to dismiss Reid's critique and his alternative position as "intuitionism," for this very term is freighted with sources of confusion. Reid's "common-sense" realism has more the character of mathematics than of psychology, and it was often to mathematical principles that Reid turned when attempting to illustrate or explain the foundations of his philosophical conclusions. Intuition may be the psychological account of how one reaches or understands the sources of the notions in question, but it is logic that dictates their necessity.

This is all obvious in mathematics. Reid was well aware of the ancient concept of "common notions" as employed by mathematicians to refer to the irreducible and necessary starting points of any mathematical system or scheme. Euclid begins with three such notions: (a) things which are equal to the same thing are also equal to one another; (b) if equals be added to equals, the wholes are equal; and (c) if equals be subtracted from equals, the remainders are equal. Analogously, for Reid, that the ideas one has are one's own is the necessary and irreducible first step in joining two ideas together or having them guide action. He makes this point in any number of places in his major works. Thus, the notion one has of one's own identity over time

> is indispensably necessary to all exercise of reason. The operations of reason, whether in action or in speculation, are made up of successive parts. The antecedent are the foundation of the consequent, and, without the conviction that the antecedent have been seen or done by me, I could have no reason to proceed to the consequent, in any speculation, or in any active project whatever. . . . From this it is evident that we must have the conviction of our own continued existence and identity, as soon as we are capable of thinking or doing anything, on account of what we have thought, or done, or suffered before; that is, as soon as we are reasonable creatures.[41]

More recently and along Reidian lines, John Hawthorne has advanced a formal argument against the Humean position that would locate mental causation in a spatiotemporal domain external to individuated mental life.[42] Hawthorne takes Hume (and "Humeanism") as contending that the causal facts operating in any region of the world are extrinsic to the local events and are to be understood as arising from what Hawthorne characterizes as "the global distribution of freely recombinable fundamental properties."[43] Hume's theory reduces causality to regularities such that, in principle, anything might be the cause of anything, just in case the freely combining external properties (Humean "objects") combine in a certain way. On this account, what is taken to be the cause of death in subregion A (a bullet entering the heart of the victim) could have a different outcome in subregion B, just in case the victim in A had been embedded in B. On this basis, Hume concludes, as Hawthorne says, "that the causal facts pertaining to a region are extrinsic to it."[44] The problem with the thesis is exposed most fully by the facts of *consciousness*. A duplicate of Hawthorne moved to any subregion that thereupon contains him now will have a resident with Hawthorne's conscious life, this being intuitively unarguable. (Note something of a "Cartesian" move here.) Thus,

> If a spatio-temporal region wholly contains me and I am conscious and some other spatio-temporal region does not contain a conscious being, then it is altogether obvious that the two regions are not duplicates. Deny this connection between consciousness and intrinsicality and it seems to me that our very handle on the notion of intrinsicality (and the coordinate notion of duplication) will be thrown into doubt. At the very least, this connection is *highly* intuitive.[45]

The warranted conclusion is that external regularities cannot duplicate Hawthorne in the new subregion, for the duplication requires the *intrinsicality* property of consciousness itself.

What both Reid and Hawthorne offer are arguments to the effect that "personal identity" is the necessary starting point, the required common notion, *indispensably necessary to all exercise of reason* and thus not subject to reason's authority. The very arguments that might be adduced to raise doubts about it presuppose on the part of the subject of such instruction the very integration of those successive parts of the thesis achieved solely by an entity whose ideas are *personally* owned. Accordingly, Hume's search for his own personal identity required those very operations of reason, functioning over time and uniquely "Hume's," in virtue of which Hume could comprehend the search as *his own*. The property of the knowledge in question has been referred to as *reflexively transparent*.[46]

Suppose one knowing oneself to desire a glass of water. Such a judgment is *reflexively transparent* in that the desire itself is, as it were, given by the occurrence of the judgment itself. In the example, the terminology refers to the *reflexive* nature of the relationship between "knowing oneself" and "desire," but it should not be thought that the knower here must engage in some sort of introspective process to determine if the desire is his own. No belief warrant is required, for it would be something of a pathological condition were one to require a warrant for believing that one knows oneself to be thirsty.[47] On this very point of pathological conditions, it is important not to neglect multiple-personality disorders and similar clinical categories as real or seeming exceptions to the integrity and continuity of personal identity.[48] A brief historical sketch of such cases provides a framework within which to test their relevance to the issue of personal identity.

Among the earliest investigators was Alfred Binet, famous for his development of IQ testing and in the closing decades of the nineteenth century, recognized as France's leading experimental psychologist. In his *On Double Consciousness*, Alfred Binet addressed this striking phenomenon, which expresses itself in the form of one or another pattern of dissociation.[49] Before summarizing his own research and that of Pierre Janet, Binet paused to give major credit to Theodule Ribot, the celebrated professor of psychology at College de France, where his teaching and writing came to influence an entire generation of French psychologists. He was, among other distinctions, founder of the French Society for Physiological Psychology. His books included up-to-date reviews of psychological research and theory in England and Germany.

The clinical-medical orientation exemplified by Ribot takes pathological conditions to be a key to the normal rather than as irrelevant exceptions to it. It is worth recalling several of the pioneers here: Philippe Pinel, his student Esquirol, and also—and even especially—Franz Gall. It was Gall's modular theory of brain function that proposed the actual mechanism by which the dissolution of the alleged unity of consciousness might be effected. His theory of separate "organs" within the brain led him to conclude that distinguishable self-consciousnesses might be thus mediated. And so we find one of Gall's scientific critics, Pierre Flourens, complaining that "consciousness tells me that I am one, and Gall insists that I am *multiple* . . . consciousness tells me that I am *free*, and Gall avers that there is no moral liberty. . . . Philosophers will talk."[50]

Within scientific and medical contexts, however, it was Gall's perspective that would win out, even if his phrenological theories ultimately went the way of the perpetual-motion machine. Long before Freud, and in a manner to which Freud would be indebted, it was this perspective that came to be widely shared among the French clinicians dealing with psychopathology. Ribot was a leader here. His writing and teaching steadfastly opposed metaphysical approaches to an understanding of the mind. He would replace these with conclusions drawn from direct clinical observation. Thus, to explain the formation of the *self* or personality, the most direct method is the clinical study of its destruction. In *The Diseases of Personality*, Ribot presents a ready alternative to metaphysics: "In seeing how the ego is dissolved, we discover how it is made."[51]

Ribot considers one of patients who complains that "each of my senses, each part of myself, as it were, is separated from me and can no longer give me sensations; it seems to me as if I never actually reach the objects that I touch."[52] Then he cites one of his own patients who claims to have been dead for two years: "Everything in me is mechanical," he says, "and is done unconsciously."[53] What Ribot concluded from such claims is that they arose from a second and entirely *independent ego* rather than a transformation of some single and putatively "unified" ego. This, Ribot claimed, is especially obvious in cases of "circular insanity," the term then in vogue for what is now called bipolar manic-depressive disorders. The manic phase expressed psychologically is the outward manifestation of the "superactivity of all the organic functions. . . . Upon what might be called the primitive and fundamental personality . . . are grafted by turns two new personalities—not only quite distinct, but wholly excluding each other."[54] In the extreme cases, such as severe dementia, there is not the successive appearance of different personalities but the actual *doubling* of the personality. Such patients regard themselves as double and will refer to "we" rather than "I." "There is no skepticism on their part as regards their state, nor do they tolerate it in others."[55]

According to Ribot, these cases, as with all the others, were to be understood as reflections of a disordered brain, a failure of physiological systems to function coherently or in a fully developed fashion, a generalization he extended both ontogenetically and phylogenetically. Harmonious individuality is the product of *evolution*; dissociative diseases are typically the result of the reverse.

Against this background, it is sufficient to say only a few words about the research and theory Alfred Binet developed based on his studies of both normal and pathological divisions or "doublings" or, indeed, *dissociations* of personality. In studies of somnambulists, of normal persons exposed to hypnotic suggestion, and of hysterical persons examined in research on "automatic writing," Binet compiled a fascinating assortment of findings all tending toward the conclusion that consciousness is divisible and separate consciousnesses can yield "the dignity of a veritable personality."[56] Binet showed that a hypnotized normal subject can be given suggestions that will mimic in their effects the phenomena of "automatic writing" displayed by hysterical patients whose writing hand is anaesthetic. Thus, hypnotism can "double" the ego in the normal person, in that coordinated and seemingly goal-directed behavior of which the actor is otherwise oblivious can be elicited. It is also possible to induce hallucinations hypnotically by suggesting the presence of a common object, for example, a key or book. The subject is told that a book is on the table, whereupon the subject reports seeing it. What is interesting about this effect, as reported and studied by Binet himself, is that the hallucinated objects behave as real objects do when the relevant perceptual responses are assessed. If the subject is looking at the hallucinated book and his eye is made to move by the exertion of pressure against it, the perceived book becomes blurred, moved, and doubled. Or if the subject moves away from the book, its apparent size decreases and its physical properties become less clear. Binet explained such findings by way of a theory of "indications" (*pointe de repere*, a "benchmark"). "The hypnotized person manages to combine the hallucinatory image with a sensation from real objects existing in the external world. Optical instruments, by modifying this real sensation, give the subject the idea of a corresponding modification in the hallucination."[57] As all real objects change size as a function of viewing distance, they provide this indication or "benchmark" to which the hallucinated object is assimilated. Note, then, how this explanation is at once a contextual theory of perception based on the assumption that any number of information-processing systems may be operative outside the scope of immediate consciousness.

SOCIAL CONSTRUCTIONISM (AGAIN)

How is all this to be understood within the framework of the unity of "self" and the continuity of personal identity? The reasonable reply is that cases be understood as the consequence of pathological conditions serving to make clear the difference between normal and diseased function. Surely no one would contend from the facts associated with adult-onset diabetes that conceptions of normal physiology now must be "relativized," nor would it be plausible to contend from the facts of color blindness that the scientific understanding of the physiology and pigment chemistry of the normal retina requires reconsideration. What was reported by Binet, Janet, and Ribot more than a century ago and later amplified by Oliver Sacks[58] demonstrates the nature of the normally functioning person and the consequences endured by that person in the event that pathological states result in the disintegration of the "self."

This is not to ignore the vigorous criticisms advanced over the centuries against all species of that "elderly Cartesianism" seen now as reviving the venerable *substance* theory of "self." The two sides in this enduring controversy were summarized with characteristic *panache* by William James:

> If, with the Spiritualists, one contend for a substantial soul, or transcendental principle of unity, one can give no positive account of what that may be. And if, with the Humians, one deny such a principle and say that the stream of passing thoughts is all, one runs against the entire common-sense of mankind, of which belief in a distinct principle of selfhood seems an integral part. Whatever solution be adopted in the pages to come, we may as well make up our minds in advance that it will fail to satisfy the majority of those to whom it is addressed.[59]

James was notoriously impatient with metaphysical abstractions or, for that matter, any claim that seemed to require nothing by way of experience or something with trumping power over all possible experience. It takes little imagination to anticipate his estimation of terms such as "self," "ego," and "consciousness," at least as these occupy some eerie realm into which mere experience can gain no access. Thus:

> I believe that "consciousness," when once it has evaporated to this estate of pure diaphaneity, is on the point of disappearing

> altogether. It is the name *of a* nonentity, and has no right to a place among first principles. Those who still cling to it are clinging to a mere echo, the faint rumor left behind by the disappearing "soul" upon the air of philosophy. . . . My thesis is that if we start with the supposition that there is only one primal stuff or material in the world, a stuff of which everything is composed, and if we call that stuff "pure experience," then knowing can easily be explained as a particular sort of relation towards one another into which portions of pure experience may enter.[60]

In this James is no "reductionist," for it is just the stream of pure experience that composes mental life, a life no less "mental" for being liberated from substances, essences, and spirits. There is, nonetheless, a real "me," a real "I," and the question for the psychologist as James saw it was to discover in what this was grounded. The identification of anyone is the identification of a *personality* that "implies the incessant presence of two elements, an objective person, known by a passing subjective Thought and recognized as continuing in time. *Hereafter let us use the words ME and I for the empirical person and the judging Thought.*"[61]

What James offers here bears a resemblance in instructive ways to traditional Buddhist thought, as has been recently shown by Jonardon Ganeri.[62] Rejected by the Buddhist is the notion of some real substantial and perduring *self* over and against what Ganeri refers to as "continuous streams of psycho-physical constituents."[63] As with James, the thesis is not reductive, but, as with James, it is one that must account for the patent relationship between references to persons and references to the very psychophysical streams constitutive of the reality in question. Ganeri makes clear a number of subtle and important distinctions among Buddhist writers on the self and the ever more subtle thought of the Buddha. Jamesian debts to Buddhistic teaching have been noted in the past.[64] What is common across the East-West divide is not just the foundational reality of the psychophysical stream but resistance to the isolation and private ownership of any part of the world and of the world as experience. James would solve the problem of reference as well as that of ontologically dangling substances by attaching mental life to the ever present passing stream of experiences, the indubitable realm of pure experience. James's solution, of course, is no solution at all. It is in proper anticipation of the reader's incredulity that he concludes his *précis* on consciousness with the conviction that, at the end of the day, the "I think" is no different from the "I breathe," and that, though consciousness "*is fictitious . . . thoughts in the concrete are fully real. But thoughts in the concrete are made of the same stuff as things are.*"[65]

Along with William James and the Buddha, there is now widespread agreement that there is no "ghost in the machine" (either because there are no ghosts or because there is no machine), and that the essential or substantial "self" or "ego" or *res cogitans* is, as James would have it, just the old "soul" of the religionists grown shamefaced. It has no ontological standing outside the "script," whether the latter be understood as a program, a network, a discourse, or the habits of a culture. The heavy philosophical work here was done by Ludwig Wittgenstein, chiefly in his critique of the private-language argument and on any and every form of solipsism. Aspects of this were addressed in chapter 2. Recall that, on Wittgenstein's understanding, to say that my experiences are "mine" and can be made known to others solely by means of my behavior is to make a claim that is either false or misleading. We think we are clear and incorrigible when saying something along the lines of "I'm the only one who knows the pain I'm feeling," but this cannot be an epistemic claim, for the pain in question is not something I've discovered or learned about through study or might know only incompletely. In point of fact, the utterance including the word "pain" is but a verbal substitute for (to use Wittgenstein's example) *groaning*.[66] Wittgenstein labored to make public and social what the Cartesian tradition appeared to reserve to the private duchy of that "thinking thing" subjected to scrutiny now for the better part of three centuries.

In one or another form, the Wittgensteinian analysis has been amplified, copied, and rehearsed by scores of philosophers, usually in a way that makes it seem as if there is no longer a problem. From the computational wing of the antidualist party we learn:

> A related reason why the mind-brain problem seems hard is that we all believe in having a Self—some sort of compact, pointlike entity that somehow knows what's happening throughout a vast and complex mind. It seems to us that this entity persists through our lives in spite of change. This feeling manifests itself when we say "I think" rather than "thinking is happening," or when we agree that "I think therefore I am" instead of "I think, therefore I change." . . . This sense of having a Self is an elaborately constructed illusion—albeit one of great and practical value.[67]

The social constructionists, with acknowledged or veiled debts not only to Wittgenstein but to Heidegger and Rorty, frame it differently: "My current project, then, is to suggest how we can begin to conceive of selfhood without the threat of solipsism and the myth of the 'self,' as a diaphanous homunculus hidden within. By the same token, we should be able to formulate a concept of agency without the myth of the will."[68]

At the superficial level, a "compact, pointlike entity" suggests a topology quite different from that of a "diaphanous homunculus." But where it matters, at the fundamental ontological level, both fall into the empty category of the illusory, the mythic, and worst of all, the "Cartesian." Nevertheless, Minsky must still acknowledge that his compact and pointlike entity has great and practical value. Even in the form of a diaphanous homunculus, the thing provides one of the bases upon which we are able to live nothing less than a given form of life for, as Harré goes on to explain, "our forms of life would be closed to people who believed otherwise."[69]

This is, to say the least, a vexing state of affairs. On the one hand, leaders of thought drawn from opposite philosophical and conceptual poles agree that the concept of the Self, as traditionally understood and defended, is mythic, illusory, and grounded in a mistake that is no less innocent for being Descartes'. At the same time, great and practical implications are tied to a loyalty to this mistaken view such that, upon abandoning it, we face the prospect of losing no less than our very form of life! The choice would seem to boil down to one between self-deception (or whatever compact and pointlike entities do to cajole themselves) and self-destruction. After all, if the Self is no more than a social and linguistic construct in the first instance, then any radical alteration in the accepted modes and terms of discourse must have the effect of installing something different where the Self once stood.

Richard Rorty, facing up to the dilemma, offers the following dispensation in an exhortatory passage that somehow fails to reassure: "We pragmatists should grasp the ethnocentric horn of this dilemma. We should say that we must, in practice, privilege our own group, even though there can be no non-circular justification for doing so. . . . We Western intellectuals should accept the fact that we have to start from where we are, and that this means that there are lots of views which we simply cannot take seriously."[70]

We are allowed to retain the myths for, after all, they are *our* myths. Confronted with arguments that would challenge such convictions, we are entitled to some measure of protective mirth and the consolation that comes from regarding our interlocutors as insincere, incoherent, or just comical. It is worth noting that the title of the essay containing this thesis is "Solidarity or Objectivity?"—alternatives no less worrisome, it would seem, than self-deception or self-destruction. Here again William James opened the door through which assaults on "substantialism" entered as a flood: "Transcendentalism is only Substantialism grown shame-faced, and the Ego only a 'cheap and nasty' edition of the soul. . . .

The Ego is simply *nothing*; as ineffectual and windy an abortion as Philosophy can show."[71]

It is to the credit of both Locke and Hume that a healthy skepticism arose about "substantial souls." Indeed, Locke and Hume were both correct in locating the autobiographical facts of a percipient's life within the arena of conscious experience and memory, whether these are taken to be (Lockean) sensations or (Humean) causal relations among the impressions. When Barbara makes a claim as to *who* she is, the claim is epistemic and as such must meet certain epistemic criteria, if only to Barbara's own satisfaction. Failing in this regard, her claim is either false or weak, and it may then be said that she simply doesn't know *who* she is. She may thus be judged amnesic, deluded, or severely confused. What she lacks is a *self-identity*. The more common instance is one in which Barbara has no doubt at all as regards who she is but, because she is in the company of total strangers, others do not have the same knowledge. She knows, because she possesses just those connected memories that knit together the unique events of a single life—one denominated "Barbara Q. Smith"—but in this room filled with strangers she is just an unknown person. What is lacking here is a *personal identity* to match up with her *self-identity*. She knows who she is but others do not. What they require are such items of information as birth certificates, Social Security numbers, histories of shared experiences, etc. Others require the sort of information possessed by Barbara's friends and family. The latter know who she is, and they would continue to know as much even if some catastrophe befell Barbara and left her without a *self-identity*. Note, then, that, as these terms might be more precisely employed, there can be a *personal identity* (the identity of a given person) where there is no *self-identity* (one's correct identification of oneself) and vice versa. However, even under conditions in which Barbara has lost her self-identity (e.g., through amnesia) and in which others (because they are strangers) cannot establish her personal identity, Barbara still knows *that* she is, which is to say that her *self* survives even when the Lockean-Humean criteria cannot be met.

No end of confusion, not to mention metaphysical eccentricities, will be spawned by the failure make distinctions among these three different entities. Constructivist theories that offer plausible accounts of the formation of one's self-identity will be wrongly judged as explanations of the *Self*. Anthropological accounts, with convincing data to the effect that socialization produces certain sorts of *persons*, supplied with defining "roles," will be misconstrued as having some bearing on the very different matter of the *Self* and will take the *self/other* boundary as culturally drawn.[72]

Selfhood is not synonymous with self-identity or personal identity. Is this because the latter terms are grounded in observable social practices and personal activities, whereas the *Self* is just that point-like illusion, diaphanous homunculus, and windy abortion dismissed by an army of savants? If William James did much to encourage such a view, he also had a remedy: "The essence of the matter to common-sense is that the past thoughts never were wild cattle, they were always owned. The Thought does not capture them, but as soon as it comes into existence it finds them already its own. How is this possible unless the Thought have a *substantial* identity with a former owner,—not a mere continuity or a resemblance . . . but a *real unity*?"[73] James's answer to this question, his "solution" to the problem, is thoughtful but not successful. He solves the problem of ownership by invoking what might be called a testation theory of the Self. As the passing thought dies, along with its current owner-Self, the next idea becomes owned by a successor-Self who enjoys title to it by a kind of automatic bequest. Accordingly, "the present judging Thought, instead of being in any way substantially or transcendentally identical with the former owner of the past self, merely inherited his 'title,' and thus stood as his legal representative now."[74]

James quickly recognized that his account does not defeat substance theories of the self. The testation theory offers no way for the very first thought to get possessed—and thus incorporated first into the estate and then into the testament—except by inhering in the very *substance*-Self the theory seeks to replace. He addresses this criticism in a long footnote, but one that so fully anticipates contemporary views as to warrant inclusion here nearly in its entirety:

> We must take care not to be duped by words. The words *I* and *me* signify nothing mysterious and unexampled—they are at bottom only names of *emphasis*; and Thought is always emphasizing something. Within a tract of space which it cognizes, it contrasts a *here* with a *there*; within a tract of time a *now* with a *then*; of a pair of things it calls one *this*, the other *that*. I and *thou*, I and *it*, are distinctions exactly on a par with these,—distinctions possible in an exclusively *objective* field of knowledge, the "I" meaning for the Thought nothing but the bodily life which it momentarily feels. The sense of my bodily existence, however obscurely recognized as such, *may* then be the absolute original of my conscious selfhood, the fundamental perception that *I am*.[75]

But even granting that indexicals such as "I and thou" are as distinct and objective as "here and there" or "now and then," it remains to be settled just what the frame of reference is for all such distinctions. The frame of reference, of course, is the percipient's own sensed location in time and space, otherwise "now," "then," "here," and "there"—which are all relative terms—can stand in relation to nothing, not even themselves! What, after all, is gained over the venerable but defective Locke-Hume account by grounding the *Cogito* in a "sense of my bodily existence," especially granting that *something* has to be the experient of this sense? And to identify that something as "Thought" leaves the conundrum precisely where it was found, for every thought must be *someone's*.

It is true—even a truism—that "I" and "me" should not be taken as in any way "mysterious and unexampled," for they refer to what *must* be indubitable to any conscious being able to have and report experiences. We might even go the extra length with James and accept the proposition that such pronouns reflect only a certain *emphasis* whereby thoughts partition their contents not only into spatiotemporal but also proprietary categories. Nevertheless, the pronomials must be referential, and to hypothesize that the "sense of my bodily existence" is the referent whenever I refer to *my* sensations or perceptions or feelings is not to challenge the substantialist thesis as much as to render it more precise. It is important to underscore the additional fact that such precision is not at the expense of the alleged "transcendental" or substantial nature of the Self, for there is nothing in the substance theory that removes the Self from any and all commerce with the affairs of the body.

SELF AND WILL: A KANTIAN AFTERWORD . . .

Kant, too, was an influential source of criticism of the "Cartesian" form of substantialism. In the "Paralogisms of Pure Reason,"[76] he grants that the soul or the "I" or the Self is a *substance,* but in an entirely nugatory fashion. The designation means only that it is distinguishable from mere predicates and determinations. All experience inheres in this "substance," but nothing else follows from the fact. "Consciousness is, indeed, that which alone makes all representations to be thoughts, and in it, therefore, as the transcendental subject, all our perceptions must be found; but beyond this logical meaning of the 'I,' we have no knowledge of the subject in itself."[77]

However, Descartes sought nothing beyond this real distinction until he attempted to trace out certain religious consequences. He meant only to establish the logical partition between mind and body and the certainty that attaches itself to knowledge of the former. Like Kant, Descartes was aware that the *empirical* dimensions of mental life and of the Self would not be revealed by a logical exercise. But we must not leave Kant's reflections on these matters at this point. As early as the preface to the second edition of his *Critique of Pure Reason* he is found clarifying the position of soul and Self in his entire system of philosophy. The two-fold manner in which really existing entities can be known—through their *appearances* and *noumenally*—applies to the Self as well. If our self-knowledge were to be confined to appearances and thus the gift of experience, it must be that the being so known is entirely determined, for the realm of observables is the contingent realm of natural causation. But this would have to be at the expense of morality itself, which "necessarily presupposes freedom (in the strictest sense) as a property of our will."[78] This leads Kant to what might be regarded as the moral form of the Cogito-argument as he considers the soul viewed from the noumenal standpoint: "My soul, viewed from the latter standpoint, cannot indeed be known by means of speculative reason (and still less through empirical observation). . . . But though I cannot *know*, I can yet *think* freedom. . . . Morality does not, indeed, require that freedom should be understood, but only that it should not contradict itself, and so should at least allow of being thought."[79]

Absent the "I," of course, there cannot be a moral being capable of thus locating itself in the realm of freedom. The *Self* becomes, as it were, the necessary and not merely grammatical requirement of morality, for only a *res cogitans* can *think freedom*. A number of possibilities arise from these considerations, not least of which is the freedom to think that the *Self* might be a *substance* after all. In any case, the more authoritative counters to such a possibility are seen to be no more compelling than the thesis itself, and certainly far less compatible with the seasonless assumptions of life beyond the seminar room.

6

EMOTION

Oh shame! Where is thy blush?
—*Hamlet*, 3.4

A persistent theme in philosophy both ancient and modern, with authorities extending as far back as the Old Testament and Homer, finds human nature to be a house divided, emotions living on the lower floors and creating such havoc that the rationality living on the upper levels is often fitful and beside itself. Indeed, the ancient Greek εκστασις is quite literally being "beside oneself." The very first word in the *Iliad* is "anger" (Μηνιν), and the entire epic charts the wages of passion's decisive victories over our better judgment. Plato's dialogues are somewhat mixed in their message here, for in them we discover that ερος is at once destructive and an aspect of the divine. Three of the dialogues, the *Symposium*, *Republic*, and *Phaedrus*, explore this in some detail, often in graphic terms.

Brief reflections on the *Phaedrus* will be useful. In it, we meet again the metaphor of the charioteer pulled by a pair of winged horses (246a6–7) as different as black and white—for they are black and white. The white one has the most honorable impulses, whereas the black horse is wild and inferior in all the noble elements. Only when reason brings the latter under control—which is to say when rationality rules animality—can a worthy destination be reached. Under the sway of the black horse's impulses, we are drawn to all varieties of excess: gluttony, wantonness, eroticism (238b).

It is chiefly in reductionistic theories that actual persons are called upon to function as if their lives include significant periods of pure rationality or pure emotionality. In a recent and highly instructive volume, Max Bennett and Peter Hacker refer to this partitioning of processes as the *mereological fallacy*, the fallacy of treating the whole as if it were but a collection of independent parts.[1] Unless the last word on rationality and emotion is reserved to the introspecting "Cartesian," and this at the cost of reductionism itself, the putatively distinct states of reason and emotion must be identified from the outside. There must be some public mark of rationality and of emotion, as well as the variants and degrees of each. The droll question arises as to just *where* each is to be found.

Where, for example, is the locus of the rational? It is trivially true that rationality is a feature or attribute inextricably bound up with the utterances, endeavors, and achievements of actual persons and, to that extent, is located wherever they are. Nonetheless, it is also the case that ascriptions of rationality are not reserved solely or even primarily to the self-reports of the persons thus uttering, endeavoring, and achieving. There are both logical canons and broad, pragmatic criteria routinely invoked to establish whether persons are "rational" in their activities. One who affirms the truth of two contradictory propositions is said to be irrational in the circumstance, "rational" here being synonymous with "logical." In matters of this sort, therefore, it would be more apt to say that the locus of "rationality" is not spatial but conceptual and found wherever one finds entities capable of framing and comprehending propositions. The Law of Contradiction is not confined to one jurisdiction or another but applies universally across all assertions claiming truth value. To ask, then, if the locus of Smith's rationality is wherever Smith's *body* happens to be would, on this account, be like asking whether the Law of Contradiction moves more quickly when Smith is cycling than when he is standing still.

Nonetheless, conceptual truths and logical canons do not begin to exhaust the domain of rationality, at least as the term connotes "reasonableness." Implicit in a rational course of action are judgments not readily subject to purely logical analysis. There is a wide gulf between the strictures of logic and the broad, open-textured realm of what is "reasonable" in a given case. The contrast is vivid and suggestive between the universal reach of propositional logic and that utterly contingent and contextual nature of the "reasonable" in actual practice. Where life is actually lived, only shifting probabilities reign, and the use of Bays' Theorem takes more time than one typically has, even if one knows anything about it.

The history of the social sciences has not been indifferent to all of this. In a foundational essay published in 1738, Daniel Bernoulli left no doubt but that the *psychological* must be regarded as central among the factors to be considered in any quantitative, scientific approach to decision making.[2] It is in the actor's subjective estimations of gain, rather than in the objective mathematical probabilities, that decisions are grounded. Persons averse to risk choose smaller gains with a higher likelihood over less probable but greater gains. However, whether one eschews risk or actually seeks it will depend on many factors, among them whether the cost of risk is significant in the life of the specific actor. One for whom a loss of one hundred dollars is negligible does not judge risk with that sense of urgency likely to beset the impoverished. Clearly, there is both an "affective" dimension to this seemingly rational calculus and also a rational moderating of the emotional investment in the various options. One begins to see that the problem here might not be one of teasing apart two distinct "processes" but coming to grips with the artificiality of the very concept of a "process."

Over the past two centuries, what Bernoulli offered as a new theory has developed into a significant chapter in the history of the social sciences.[3] Considered broadly, research in this area has established not simply the so-called subjectivity of decision making but also the highly *personalized* character of the subjectivity itself. Just how a decision is likely to affect the specific actor—how it figures in that specific life under those specific circumstances—is decisive in any attempt either to predict or plausibly explain the decision to choose a given course of action. What matters centrally are not gains and losses neutrally or statistically considered, but *interests*, as these bear on actual persons. And where such significant interests are at stake there inevitably will be found *emotions*. To take an interest, to take a great interest, is to invest a situation with what is personal, meaningful, and *felt*. To the extent, then, that the subject of so-called rational decision making must incorporate this personal equation, which itself includes the actor's appraisal of the situation and its bearing on significant personal interests, it is clear that what is called *rational decision making* has already been shaped and colored by affective contents.

To give this truism something of an axiomatic character, it may be said that:

(a) A decision is rational when it bears reasonably on the interests of the actor.

(b) The interests of the actor include those conditions, possibilities, and aspirations about which the actor has non-negligible feelings.

(c) Actions clearly at variance with the protection or advancement of just those interests for which the actor has such feelings are, in the circumstance, nonrational, perhaps irrational.

In this more axiomatic form, one truism incorporates yet another: to the extent that a decision's rational standing depends on the coherent relationship between the felt interest and the decided course of action, the element of rational deliberation is an essential feature of the overall situation. The point is worth repeating and illustrating. It is not sufficient that there be reasonable conduct and also a realistically and keenly felt interest. Mary is keenly interested in getting to the bank, which is ten miles away, before it closes. To reach the bank in time requires driving up to the legally maximum speed. But school is letting out, there are children running hither and thither, and safety would call for a speed less than the lawful maximum. Nonetheless, Mary drives at the lawful maximum. Here we have a realistic interest, keenly felt, which, *ceteris paribus*, makes driving with dispatch reasonable. But it is *unreasonable* to be merely "lawful" in the circumstance. The combination of having a good reason and having a keenly felt interest is insufficient, for what is lacking are the moderating influences of still other reasons reached through deliberation and with the aid of such signal interests as the desire not to harm.

The point, of course, is that, in actual life, not only is the emotional an integral feature of the rational, but the rational is omnipresent in all attempts to preserve and promote one's significant interests. Interests may be and often are in conflict. It is not inevitable that the greater will prevail in setting a course of action, nor is it invariably the case that one has full comprehension of one's interests. Indeed, if action were no more than a response to feelings, there would be little need for rational deliberation at all. Part of the whole point of such deliberation is weighing the merits of those feelings that incline one to act.

Further refinements of these terms and concepts are required, beginning with an attempt to clarify the very meaning of the word *emotion* as it has come to be used in psychology and philosophy. Use has been profligate, and it is necessary to impose some order. Just what is an "emotion," in what sense is it "felt," and how it is related to one's "interests"? How has philosophy come to understand and approach these questions, and what guidance has been forthcoming from science?

Just what an emotion is remains arguable, but research and theory have combined to set down seven basic emotions found in a large number and variety of cultures. The authoritative work in this area is that of Paul Ekman.[4] His evolutionary theory of emotion is indebted to Darwin's foundational work on the same subject.[5] Ekman's study of facial expressions and his detailed examination of the pattern of activation of the facial musculature have led him to identify as the seven basic emotions those of anger, sadness, fear, surprise, disgust, contempt, and happiness. The speed with which arousing conditions result in the characteristic expressions is often in the range of microseconds, indicating that an undeliberated and presumably noncognitive adaptive mechanism is at work. Nonetheless, it is obvious that what excites anger or contempt or happiness will depend on factors having nothing to do with biological functions, even if the expression of these sentiments is stereotypical at the level of the facial muscles. Beyond this, there is the difference between such basic emotions and what are commonly called *moods*, in that the latter are not readily or reliably associated with specific and causally effective conditions. Typically, one is angered by the utterances or actions of another, surprised by the appearance of an unexpected stimulus, saddened by the loss of a loved one, etc. But there are also broad, more or less atmospheric states of mind: a general sense of well being or of impending trouble or of deep disappointment. Those reporting such states often cannot point to a specific cause or condition and surely would not maintain a fixed facial expression throughout the duration of the mood, which might be a fraction of a lifetime.

Jaak Panksepp's *The Foundations of Human and Animal Emotions*,[6] which appears in the *Series in Affective Science*, defends the evolutionary and neuroscience model of emotion rigorously but with a sensible respect for the complexity of the subject, not to mention the ethical implications arising from the very assumptions on which such research is based.[7] What the new "affective science" is able to add to the Darwinian account is a wealth of information drawn from neurochemical, physiological, and anatomical studies of the nervous system. Panksepp draws upon these studies and concludes that a rich and varied emotional life arises from the functions of the nervous system and serves the purposes of the individual organism and the species as a whole.

Throughout his book, Panksepp gives his chapters and major sections titles suggesting that the brain (somehow) creates joy, rage, loneliness, grief. He writes in a contemporary genre and uses jargon that has been adopted even by the popular press. Most of what was once attributed to persons is now routinely attributed to brains, as if the attribution itself settled the most fundamental questions. It should be clear, however, that if sentiments and feelings and moods are features of conscious life that are unlike *anything* in the domain of physics, little is gained by moving the problem of explanation from the person to the person's brain. Similarly, so-called *functionalist* explanations, as old as ancient Greek philosophy and fully canonized by Darwin and his disciples in the nineteenth century, say nothing whatever about the phenomenology of emotional life. To learn (as if it were at all surprising) that fear and rage have survival value is not to account for the emotional feelings. It is instead to locate them within the larger framework of such self-preservative functions as digestion and respiration, neither of which requires consciousness. There is no obvious reason why a creature confronted by a dangerous set of conditions could not be designed to take flight without having any feelings whatever. Nor is it obvious that such activities as mating or gathering in numbers or sheltering a litter require an emotional component or are even more effective as a result of emotion.

Consider again Paul Ekman's influential research and theory. Extensive studies of different cultures turn up not only stereotypical facial and postural signs of emotions but words for these that appear to translate readily from one culture to another. What counts as "happiness" in one culture is expressed pretty much as it is in another and is referred to by words and phrases more or less alike. It would not stretch credulity to assume that the same is the case with human cognitive resources and with the biological basis of motives and desires. Taking all this into account, there is an undeniable fact that sits uncomfortably amidst this sameness: the immense, qualitatively distinct, historically enduring differences among the world's identifiable cultures. Creatures possessing the same fundamental cognitive, motivational, and emotional "structures" nonetheless create and maintain widely differing systems of law, approaches to life, attitudes toward each other, religious and political allegiances, and aesthetic preferences. Compared with any other species, human intraspecific variations are simply orders of magnitude greater than what is found in the larger animal economy. The variations are sometimes greater between neighboring tribes (with virtually identical environmental conditions and resources) than between geographically dispersed nations.

A finding of this sort, were it at the level of musculoskeletal formations or metabolic physiology, would suggest nothing less than epidemic transformations in the basic physiological functions of entire human collectives. Evolutionary theory could survive such findings only by explaining them in terms of mutation and pathology. With a figurative shrug, we are inclined to say simply that there is "cultural diversity," as if the adjective somehow removed the difficulty. The difficulty, however, arises from the very power of culture *qua* culture to moderate, modify, and transform the manner in which affective and cognitive resources are deployed, to such an extent as to render specious any theory casting biology as determinative. When Hamlet asks of Gertrude, "Oh shame! Where is thy blush?" the question is not about facial expression and coloration but about what Hamlet takes to be his mother's moral blindness. Were she to attempt to explain her actions in the language of brain chemistry and evolutionary theory, the audience would assume that this was Shakespeare being ever so cleverly droll. If, on the other hand, she were to claim that, all in all, the cultural values then prevailing permitted and even encouraged a widowed queen to marry her brother-in-law, the murderer of her husband, and now the new king, all within a month's time, Hamlet would surely have condemned the custom and the queen who honored it. The point, of course, is that even the powerful forces of culture are themselves subject to enervation by one person's passionate grief or powers of moral discernment or even mere quirkiness.

Such considerations were at the center of ancient thought, particularly when Stoic philosophy examined the relationship between emotion and reason. Two excellent treatments of the subject have appeared recently, making clear that one who has the feelings that qualify as emotional must be satisfied that these are the right feelings for the prevailing conditions.[8] It is not uncommon for someone to complain of feelings that "don't seem right" or are "far too intense" or "persist beyond all reasonable bounds." Stoic commentators were not inclined to regard any feeling as an emotion unless accompanied by the requisite rational judgment of the object or event to which the feelings were responsive. The relationship between feeling and fact or feeling and judgment is, on this understanding, subject to appraisal. If Gertrude had said to Hamlet, "Your uncle, Claudius, killed your father and I condemn him for that. Therefore I am really deeply proud soon to be his wife," there would be a clear violation of the very logic of emotion; viz., an object of contempt causally yielding pride through association. It wouldn't matter if, in the circumstance, the brain functioned normally, its happiness and contempt "centers" each displaying heightened activation. Indeed, whatever the neurophysiological facts might be, Gertrude's two sentences in the circumstance would be unintelligible to Hamlet and even to herself.

Rehearsed here is not the trite fact that human emotions are complex, contextual, individuated, etc. Rather, a sober question arises from these truisms as to just how much progress in the matter of human emotion can be gained by way of neurobiology and evolutionary theory. Progress takes different forms, of course. Further research will certainly add to the existing fund of correlations between rather stereotypic expressions of emotion and specific regions of the brain and chemical pathways within it. But progress toward ever more complete and convincing explanation is a different matter. It is as if one wished to explain the appearance, between 1150 and 1300 a.d., of more than one hundred Gothic cathedrals in Europe and Great Britain, only to be told that evolutionary neurobiological processes are the grounding of an instinctual need for shelter. If this is the right sort of explanation, then it would be reasonable to place along the same continuum nests, heaps, mounds, holes, shells, leaves, townhouses, and Chartres. This is not convincing.

Emotions are "felt," but so are proprioceptive and somesthetic sensations. One is aware of one's sensations and feelings, or, better, one *senses* and *feels*. To be thus conscious is to be conscious of *something*. Although dictionaries cannot settle philosophical disputes, how words are used must not be ignored. In ordinary discourse, the word "perception" is used when that of which one is conscious is an object in the external world. When the object of consciousness is not external but internal to the percipient, it is customary to refer to a "sensation." Again, speaking in the normal idiom, it is customary to say that the objects of perception reach consciousness through the medium of specific sense organs and are recognized as being located in a space external to the percipient. "Emotion," on this understanding, is a species of sensation rather than perception, but it is, unlike sensations, not associated with specific parts of the body, as in instances of pain. Typically, the sensations that generate emotion terms are not responsive to the specific ameliorative that works so effectively on, for example, hunger, thirst, and pain. Accordingly, and in the Aristotelian sense, emotions are more aptly understood as both *dispositional* and *evaluative*. It would be entirely permissible to speak of thirst, for example, as "dispositional," in the sense that thirst disposes one to seek and consume liquids. However, thirst as such is not a state dependent in any way on the evaluations of the percipient. It would be hopelessly academic for someone to judge whether his or her thirst were reasonable, proper, meritorious, justified. But feelings of anger, love, confidence, melancholy, pride, shame, grief, joy, contentment—these are feelings inextricably bound up with evaluations of oneself, of others, of the world, of oneself in relation to the world and to others, and so on. One may be wrongful in one's anger and blameworthy in one's love, but one cannot be wrongful and blameworthy for a toothache or for pangs of hunger. In a sense, then, the fact that emotions are subject both to personal and to social appraisal of a moral sort renders them unique among sensations and locates them in the rational realm of the person and of other persons as well.

More needs to be said on this salient point. In the case of emotions, the dispositional property depends on the evaluative, and it is this feature of emotional life that renders the person subject to moral praise or blame. It is in the identification of those conditions toward which one is disposed angrily or lovingly or proudly or sadly that one becomes recognizable as a certain kind of person, as having a certain kind of *personality*. Taking pleasure in the suffering of others, being angered by the achievements of others, loving toward the wicked and hateful toward the virtuous—these are emotions that qualify one for censure and rebuke. Note that it is not the emotion per se that qualifies one for moral praise or blame, but the class of persons and conditions toward which one is disposed to have such feelings.

Proposed here is a form of *cognitivism* in the matter of emotion—or, for that matter, a form of *emotivism* in the matter of reason, as I will suggest at the end of this chapter. It is surely different from "cognitivist" theories of emotion as set forth early by Anthony Kenny and influentially developed by Robert Solomon.[9] Against such theories, one might agree with Peter Goldie (*contra* Damasio; *vide infra*) to the effect that "emotions" and "feelings" should not have been artificially separated by analytical philosophy.[10] It is not at all obvious that "feelings," as distinct from "emotions," are in any way noncognitive, except when they are actually passive sensations now ambiguously labeled as "feelings." Consider the feeling of anxiety. Perhaps it cannot be cognitively connected with anything specific or real in the world or life of the person. In such a case, the person then judges the feeling to be unreasonable, though no less aversive for that. The point is that such "feelings" serve as a warrant for rational and even moral appraisal in ways not applicable to, for example, feeling a cloth to be rough or a glass to be warm.

Arguments between cognitivists and noncognitivists have been addressed and enlarged by research and theory in the brain sciences. Antonio Damasio has been particularly influential, partly by recovering William James's venerable theory according to which emotion is a strictly corporeal affair.[11] Damasio makes the Jamesian distinction between emotional states and the awareness of such states, dubbing the latter *feelings*.[12] Feelings are experienced, the underlying conditions being physiological and neurochemical. The latter, as composite states, constitute emotion as such. Understood in these terms, what is *felt* is the emotion. For this to take place, it is necessary that there be the function or property of what Damasio calls *core consciousness*.[13] This consciousness evolves from more elementary degrees of consciousness. At a given stage of evolution, the creature not only has the feelings but knows it. Thus, somatic representations of emotion within the brain, when "accompanied, one instant later, by a sense of self in the act of knowing . . . they become conscious. They are, in a true sense, feelings of feelings."[14] It is not enough to have feelings; one must also be able to reflect on the feelings if these are to be accorded the status of what traditionally have been understood to be emotions.

It might have been expected that the author of a work titled *Descartes' Error*[15] might be especially wary of the pitfalls of "Cartesianism." If in fact there can be no authentic and full-fledged *feelings* without a consciousness that rises above the "protoconsciousness" of simpler organisms, then a *res cogitans* must be present for those merely brain-body states to have any standing in the domain of the emotions as traditionally understood and described. Damasio is not unique in seeking to avoid these pitfalls by replacing the dual-substance theory with the scientifically correct *representations*, but these surely are as jury-rigged from the first. It makes sense to say that "tiger" is the abstract lexical representation of a large, striped, quadrupedal mammal found chiefly in India. It is far less clear to propose that a set of discharge patterns or chemical changes are the *representation* of a *feeling-plus-consciousness-of-it*. Moreover, if feelings in the sense in which Damasio would have them understood call for emotions to be somehow *knowingly* felt, a perilous regress begins to infect the entire account.

At a more general level, the arguments developed by Damasio are cogent because they are tied to clinical observations of human beings whose brain pathologies have been carefully studied. It is a bromide to say that in this the laboratory and the clinic function together, research on nonhuman animals permitting tests of hypotheses based on clinical observation. This may serve as some sort of justification for the sorts of procedures inflicted on animals, but it is less than credible as an account of the history of the current status of attempts to understand emotion. It is worth repeating that the intensity, range, and complexity of affective states achieved by way of standard laboratory methods simply falls too far short of lived life to be credible as a model of reality.

If it was, indeed, Descartes' error to have pulled rationality out of the body and installed it in a massless and unextended somewhere, the error did not include the severing of feelings and real interests from cognitive and rational life. It was precisely because of the power of emotion and desire that Descartes sought asylum in the purely rational contexts of mathematics and logic. At the risk (again) of seeming to be an apologist for Descartes' theory of mind and *relata*, I must draw attention to the detailed and scientific analysis of human emotion Descartes developed in his last philosophical work, *The Passions of the Soul*, which he intended to be an entirely naturalistic-scientific account.[16] In Article 27, he declares that all the passions are tied directly to motions of the "animal spirits" and thus are corporeal in origin. If, according to the division advocated by Damasio, we refer to these as emotions, the question then is just how these rise to the level of feelings. Descartes obviously saw the need for such a transition and it was for this that he (infamously) chose the pineal gland. It would not be stretching the rationale to say that for Descartes the "representation" of bodily states that would come to be consciously felt was the work of the pineal gland, but that the actual feelings, as the soul (mind) would know these, was incorporeal. Even Descartes' fundamental passions, set down in Article 69, match up reasonably well with the contemporary taxonomies: love, hatred, wonder, desire, joy, sadness. These are conditions that impel action in such ways as to promote survival and other benefits. As he says: "The function of all the passions consists solely in this, that they dispose our soul to want the things which nature deems useful . . . and the same agitation of the spirits which normally causes the passions also disposes the body to make movements which help to attain these things."[17] Emotion as such cannot be its own judge and jury. It is by way of rationality that the emotions are mastered, diminishing if not eliminating the evils arising from their profligate expression. But these same emotions ground nothing less than human happiness, in a word, "all the good and evil of this life."[18]

Damasio's common-sense position (when not encumbered by the speciously validating methods and jargon of "brain science") is that what is ordinarily taken to be rational activity is typically laced with affective ingredients. He is surely on the right track here, though one that has been a conduit for similar positions advanced over the ages and well rehearsed long before Descartes. If the point needs repeating for contemporary audiences, it is made with greater clarity and relevance by Donald Norman than by "neuroscience," though, again, it is a commonplace.[19] The alleged error committed by Descartes was the disembodiment of reason. As discussed earlier in chapter 2, this maneuver arose from the epistemological divide—the still acknowledged "gap"—separating the never more than probable claims arising from world-body interactions and the indubitable claims of the *Cogito*.

As for the debates between cognitivist and noncognitivist, interesting though they are, they are finally beside the main points under consideration, which pertain not to the fundamental character of "an emotion" but to a sound psychological and scientific approach to actual life. And on this point it is essential to comprehend the inextricable connection between the rational and affective features of life whenever life itself summons one to consequential actions. Moreover, to describe the position advocated here as "Aristotelian" can be misleading, for Aristotle's treatment of the emotions ranges over many of his works and differs somewhat among them. Of those works, the position advocated here comports most closely with his treatise on *Rhetoric*, for it is in this work, a work in "applied psychology," that the student is taught how to recognize, anticipate, and influence the emotions of others. If the rhetorical performance is to be successful, the speaker must know how others perceive a situation and are inclined to react to it, and how these perceptions and inclinations might be accentuated or revised. In light of the aim of rhetoric, the distinction between "feelings" and "emotions" might rise no higher than the level of a quibble about words. The aim of rhetoric is to excite action by means of persuasion. To be effective, it must be based on a valid psychology of action, a valid psychology of motivation and emotion, and a valid psychology of the states and conditions that move persons. In the end, the rhetorical achievement is one that supplies a compelling reason for action, a reason made compelling by connecting the action to the auditors' significant interests, now shown to be in jeopardy.

However, it is not enough for rhetoric to arouse feelings and excite action. Rather, it must establish the attitudes and perceptions, the judgments and motives capable of sustaining activity over long periods of time and under unpredictable and changing circumstances. Put another way, rhetoric succeeds when it creates within the actor just those states of belief, desire, and feeling that are self-sustaining, abiding until the larger aims have been achieved.

Against these several considerations, it is useful to weigh conventional approaches to the "psychology of emotion" as practiced in laboratories and featured in contemporary textbooks. Nearly any one of them will do and, with apologies, it may also be said that nearly any laboratory will do for purposes of comparison. Perhaps the right place to begin, owing to its current celebrity, is the "neuroscience" laboratory. The rationale on which efforts are expended in this domain is as follows:

1. First, "emotion" is essential to survival, for only through appropriate states of fear, anger, sexual appetite, and so on might the individual organism and the species itself survive.

2. Emotion, then, is the result of evolutionary forces shaping the adaptive potential of species.

3. Mediated at the level of the primitive nervous system, emotionality is basically instinctual and reflexive, subject to modification through experience in the more advanced species.

4. Among the advanced species, instinct is supplemented by perceptual and cognitive processes by which the emotions are moderated and refined.

This straightforward evolutionary account, now more or less official, may very well be correct, in much the same way as would be the evolutionary account of the appearance of finger-thumb oppression or erect bipedal ambulation. But it is assuredly *not* an account of anyone's emotions, any more than the evolution of finger-thumb oppression accounts for what will be written with a fountain pen or that erect bipedal ambulation accounts for last week's stroll in Central Park. More generally, scientific theories regarding the evolution of mechanisms and processes are not shorthand and "objective" accounts of lived lives, for they are not accounts of anyone's life at all.

To put the matter in a trite but economical way, every life is someone's, and so too is every feeling and every reason for acting. Accordingly, it is mistaken to assume that the combination of evolutionary biology and neuroscience offers a preferred path to a richer understanding of emotion. To the extent that evolutionary biology and neuroscience provide no more than generalizations regarding the appearance and functioning of processes and systems more or less universally distributed within a gene pool, then to that extent they can offer no distinguishing explanation or account of a given person, a given life, a given emotion. The reason is obvious: persons find themselves in unanticipated circumstances, left with choices often fewer and less appealing than desired, uncertain as to the consequence of selecting one over another, and then surprised by the turn of events loosely determined by the choice itself. The circumstances themselves are only trivially physical; what counts is how they are perceived and understood, for it is in these judgments that one concludes that this or that interest hangs in the balance. It is, then, in these judgments that one or another sentiment is aroused: fear or anxiety, perplexity or hopefulness, vigilant attentiveness or stoic resignation—the list is long.

In addition to these momentary or episodic encounters, persons also live under largely abiding conditions that may be frustrating, irksome, challenging, tedious, daunting, dangerous, supportive, fortunate, nurturing, loving. These are terms that often describe a given life on the whole, though perturbations of one or another sort do occur. Under unusual circumstances, the emotions or feelings aroused by the dangerous, the perplexing, the supportive, and so on might be intense, nearly disabling. The circumstance that arouses sorrow might worsen to a degree that leads to an enduring depression. The point to be made here is that in actual life the duration and intensity of emotions, not to mention their variety and the conditions to which they are personally attached, render utterly futile any hope of realistic laboratory simulations.

To return to the accounts advanced by evolutionary biology and neuroscience, one may ask not whether they stand as good explanations but whether they explain at all. It is a question raised acutely by Hilary Rose and Stephen Rose in their controversial *Alas—Poor Darwin* (2000). Their conclusion is that evolutionary psychology is more ideology than science, but this is not the question under consideration here. Rather, the question at issue pertains to *explanation* and to the criteria rightly applied in determining whether an explanation is a good one. Illustratively, consider the anger Smith feels toward Jones. It is in response to lies that Jones has told at the expense of Smith's good name, lies tied to the well-known envy Jones has long harbored owing to Smith's achievements. Smith's anger expresses itself in various ways, mostly disguised either by humor or condescension. Now, if Smith is asked to account for the hostility he feels, it is obvious that he will make no reference either to the theory of evolution or to events taking place in his limbic system. Indeed, ages before there was any awareness of the theory of evolution or the existence of limbic systems, persons were angered by injustice and could give good accounts of just what they took to be the injustice. Even today's evolutionary biologist and neuroscientist—should they both be well acquainted with Smith and Jones—would surely not account for the tense relationship between them in terms of evolutionary theory or brain function.

There is no surprise here, for the very concept of an "account" more or less rules out the sorts of mechanistic, causal explanations grounded in scientific laws and principles. An "account" of instances such as human relationships is something of a story or narrative in which one discovers how the participants understand the conditions facing them and how these understandings give rise to certain sentiments, desires, and decisions to act. The story told, the listener now can make sense of the events through an essentially empathic and sympathetic process. Thus: "Were I faced with this situation, and were I to perceive it the way Smith did, then surely I, too, would be angry with Jones." How odd it would be to substitute for this something like: "Had my limbic system acted as Smith's is acting, I, too, would be angry with Jones." This version fails not because it is a less informing account but because it is the wrong sort of account and thus, in the circumstance, is not an account at all. It is as if one asked what it is that makes that building a "dwelling" and were told, "the bricks." True enough, the building in question is a brick building. But the concept of a "dwelling" reaches something different from material composition. And Smith's anger over injustice—an anger that may well require any number of evolved physiological processes—reaches something not reachable by way of physiology at the level of explanation.

The reciprocal patterns of influence between the rational and the emotional, between "reasons" and "feelings," are personalized in a still more intimate sense. An important and even signal difference between emotional as distinct from other kinds of feelings is that the former are authoritatively identified and explained by the person harboring the emotion. In these respects, accounting for emotional feelings is different from accounting for the behavior that might be impelled by them. Thus, in accounting for Smith's behavior, it is surely not obvious that Smith has the last word. There seems to be enough in psychoanalytic theory and in the arts of self-deception to leave room for the possibility that Smith is entirely unaware of the cause of his pointed humor and condescension toward Jones. Note, however, that in that case it is no longer anger that needs to be explained—for there is no anger where there is no one who is angry, and, *contra* Freud, one cannot be "unconsciously" angry. The fact that one might often be unable to account correctly for the causes of one's behavior has, alas, no bearing on just who has the last word in the matter of an emotional feeling. To know if Smith is angry, the authoritative evidence is supplied by Smith alone. This special epistemic authority does not extend to any theoretical position Smith might take regarding causation. That he traces his anger to evolutionary forces or to discharge patterns in the limbic system or, for that matter, to the astrological sign under which he was born simply locates Smith within that vast community of earnest theorists.

There are, to be sure and as noted above, "feelings" without reasons for them, some so intense as to warrant therapy or medication. There is so-called free-floating anxiety or the sense of impending doom or an overwhelming but seemingly unoccasioned depression. Are such conditions to be regarded as authentic *emotions*, and, if so, do they not raise doubts about cognitivism itself? Questions of this sort threaten to hold thought hostage to a word, but there can be little clarity of thought where there is ambiguity of expression. At issue is whether one or another feeling, emotional in tone and quality, is one for which a reasonable account can be given by the experient. Where the feeling is strong and abiding and where it is endured by one who has no explanation for it, it seems right to assume that it is *caused* and caused as other sensations are, such as pain and tickle. Reduced to a maxim, the thesis is "*No emotion without a reason.*" The direct implication is that the right sort of account of emotions is the first-person narrative, that slice of autobiography in which feelings are intelligibly tied to context, belief, judgment, and perception. Again, there would seem to be little room for improvements to such accounts coming by way of neurophysiology or evolution. Where persons themselves are unable to find a reason for the feelings, and where there is no account that satisfies the person actually having the feeling, the only "account" is the *causal* account provided by way of the determinative laws of science.

There are criticisms of cognitivism and these are not slight. One worrisomely counterintuitive implication arising from it is that neither infants nor any nonverbal creature can be said to have emotions as such. This is surely the result for the brands of cognitivism proposed the Stoics of old and by such of our contemporaries as Tony Kenny, Robert Solomon, Justin Oakley, and others.[20] For a state to qualify as emotional, it must be accompanied by one or another "propositional attitude," such as a belief. For one to be angry with Sue, one must believe that Sue has wronged one, etc. To feel love for Hermione, one must desire to be in Hermione's company, believe her to be worthy, etc. And, just in case desires and beliefs of this nature are required of emotions, the counterintuitive consequence that seems to follow is that nonhuman animals and preverbal children are devoid of emotion. Interesting though it is, the debate between cognitivist and noncognitivist is finally beside the point under consideration. The point under consideration pertains not to the fundamental character of "an emotion," which may be little more than a word, but to a sound psychological and scientific approach to actual life and its competent interpretation. The question, then, of whether an infant or a dog or cat should be an object of emotional ascriptions would be provisionally settled by examining the extent to which behavior suggests some sort of deliberation prompted by some recognizable or plausibly inferred interest.

Having long enjoyed the company of dogs and cats—a benefit apparently not conferred on Descartes—I am quite comfortable with the judgment that they do make decisions and that at least some of their decisions are best understood as proceeding from expectations that qualify as beliefs. That, in addition to all this, they would somehow have to frame propositions in order to be joyful in seeing the stick in the air or the ball of wool rolling on the carpet seems to me something of a merely academic scruple designed to please journal editors.

Is my position finally "Aristotelian"? This can be misleading. Aristotle's treatment of the emotions ranges over many of his works and differs somewhat among them. What I share with Aristotle is his realism regarding the facts of nature, including human nature, and his commitment to understand complex phenomena in terms of the ends they realize or represent or suggest. He was not given to the *mereological fallacy* and would, I suspect, not be won over to accounts of emotion reduced to events in any organ, including the brain. It is safe to say that he would still insist that biological processes are not actual accounts of lived life, for they are not accounts of anyone's life at all. Put another way, every life is someone's, and so too is every feeling and every reason for acting. Accordingly, it is mistaken to assume that the combination of evolutionary biology and neuroscience offers a preferred path to a richer understanding of emotion. To the extent that evolutionary biology and neuroscience provide no more than generalizations regarding the appearance and functioning of processes and systems more or less universally distributed within a gene pool, then to that extent they can offer no distinguishing explanation or account of a given person, a given life, a given emotion. The reason is obvious: persons find themselves in unanticipated circumstances, left with choices often fewer and less appealing than desired, uncertain as to the consequence of selecting one over another, and then surprised by the turn of events loosely determined by the choice itself. The circumstances themselves are only trivially physical: what counts is how they are perceived and understood, for it is in these judgments that one concludes that this or that interest hangs in the balance.

Does all this again rule out infants and furry or feathered beings? Sheltered from our own naked vanity, we may readily grant to the balance of the animal kingdom states and feelings by which the pursuit of interests becomes intelligible, but without requiring the states, feelings, or pursuits to be akin to our own. If this seems to disconnect us from the rest of nature, not to worry: it is only a theory that is threatened by the gambit. If Jack is angered by what Jill finds merely silly, and if France takes umbrage with what Japan calls "champagne," surely it is possible for us to be thrilled by what bores a fish or causes the wren to find another branch. The uniformities of nature, such as they are, are to be found, not legislated.

In speaking of the rational and reasonable, of having a good reason to feel a certain way or to be disposed to feel a certain way, more must now be said about the rational dimension itself. Consulting the literature of the past two or three decades (not to mention the history of human life on Earth), there seems to be no doubt but that rationality is rather episodic in human affairs. The experimental literature alone leaves little room for pride in this regard.

Thomas Gilovich devotes an entire volume to presenting telling evidence against such claims as might be advanced in favor of *Homo sapiens sapiens*. To that second "*sapiens*," the title of Gilovich's book is a sobering reply: *How We Know What Isn't So*.[21] A single datum makes the case: 94 percent of American professors regard themselves as professionally superior to their colleagues! When this literature is scanned, however, it is too easy to miss the forest for the trees. Amidst studies of attribution errors, confirmation biases, availability heuristics, peer-pressure effects, and the like, one discovers refined, progressive, systematic attempts to identify challenges to rationality and those processes tending to subvert it. The literature on the whole serves as a species of what might be called applied epistemology—perhaps "*clinical epistemology*"—serving as proof not of the essential irrationality of human judgment but of the self-correcting powers of that very judgment. From the fact that human beings are subject to various diseases it follows not that human life is "essentially" unhealthy; rather, that certain habits and conditions conspire with the frailty of the flesh to pervert or corrupt what is an otherwise healthy system.

So, too, with research into the peculiar and often surprising means by which an otherwise reasonable creature reaches conclusions flagrantly at odds with the very logic of the case. But then what of the very logic of the case? As it happens, the logical exercises of the schoolroom ill prepare one for the sorts of decisions that must be made in actual life. Except in the schoolroom, there is little opportunity for the formation of precise and serviceable habits. Nonetheless, the schoolroom, the library, and the logician can be located when needed, as can the emergency room, the antibiotic, and the competent physician. In the domains in which what we take to be our most significant interests are at stake, even more than the resources of the schoolroom and the textbook logician are summoned. Consider the rule of law, fair trial procedures, rules of evidence. In any regime respectful of the dignity and liberty of the person, high barriers are erected to protect the otherwise powerless defendant from the otherwise unopposable powers of the state. Proof beyond reasonable doubt now works to tame the enthusiasms of an offended people; an adversarial dialectic is engaged to impose burdens on prejudice and to instill useful doubts, ambiguities, competing and contradicting narratives. All this rational apparatus is employed, finally, in that spirit of justice, that *fidelity* to law, which lives within the person as a noble sentiment. Through the civilizing influences of culture, through the civilizing influences of saints and heroes, the ordinary citizen comes to comprehend more fully what otherwise might be an unnamed sentiment, namely that *mercy is the perfection of justice*, here the relationship between the rational and emotional being nearly geometric.

It will be a great gain to psychology to begin to take ever more seriously the busy world of persons, striving to realize in fact what is desired in thought and urged upon them by their stronger sentiments. By taking this world seriously I mean something different from strident stipulations and ipsedixits. The seriousness I have in mind requires giving the benefit of doubt to the conventional wisdom of folk psychology rather than the conventions of academic disciplines oddly regarded as having greater authenticity or validity. Philosophers and psychologists now form an enlarged circle of specialists arguing about the relative power or place to be occupied by "the emotions" and "reason" in "determining" behavior. The broad division between "cognitivists" and "noncognitivists" is not on the question of the ontological distinctness of reasons and emotions but on which has ultimate causal or trumping power: which is the better predictor. However, what daily actual life makes clear is that those of our undertakings that we regard as worthy of philosophical or psychological interest will not admit of such partitioning or even of such a distinct ontology. Having a good reason to act is to be disposed in ways invariably described in terms of confidence, conviction, satisfaction, hopefulness—terms of "feeling" and, yes, terms of judgment. To think of these terms, however, as referring to partitionable features of mental life is to think that one can pull out of the cup what is hot, what is sweet, what is liquid, and what is brown, all the while accounting for "hot chocolate."

It will be a great gain just in case this more realistic inquiry begins with reasonable assumptions about just what goes into a good reason for action and just what goes into a truly warranted set of beliefs, desires, and sentiments. Unless the disciplines of both philosophy and psychology are prepared to take a stand on the sorts of life, at once personal, social, and civic, likely to preserve and refine both the rational and the emotional, the study of each and either is likely to be arid and actuarial.

7

MOTIVES, DESIRES, AND FULFILLMENT

"I will begin by stating . . . that it is not profitable for us at present to do moral philosophy; that should be laid aside at any rate until we have an adequate philosophy of psychology, in which we are conspicuously lacking."[1] Elizabeth Anscombe reached this conclusion about sixty years ago. It is still timely, for it remains the case that academic ("scientific") psychology is not now an adequate source for those who would understand the nature and the grounding of motives and desires and how these enter into considerations of a fulfilling life. It was this same essay that helped to revive interest in "virtue ethics," chiefly by drawing attention to the weaknesses of the dominant schools of ethics circa 1958; viz., Kant's deontological moral theory and various versions of utilitarianism.[2] What Anscombe found sorely lacking in the dominant schools of moral thought were elements that matched up with the sorts of activities and decisions actual persons make in contexts they understand to be morally weighty, or, simply, weighty. Her attention to what now falls under the heading "virtue ethics" was based on the recognition that ancient philosophers, chiefly Aristotle, had approached such issues with a commendable dose of realism. When Aristotle turns to the immensely important question of just what it is that impels persons to act decisively, he turns not to the biological basis of hunger and thirst or, in today's terms, to little bright patches on the fMRI. Rather, he turns to rhetoric. I will follow his example and begin with his own still authoritative examination of the subject. As this is a subtle and somewhat variegated set of conceptual connections, I should set up the analysis in outline form.

One of the central issues in philosophy of mind is, of course, that of *mental causation*, which has been met earlier in these pages. The "Cartesian" ghost somehow must induce and guide the movements of the "Cartesian" machine, though the two are substantially different. To speak of "mental" causation, however, is to speak in the language of common sense, once the metaphysical issues are provisionally set aside, for ordinary persons (as well as philosophers in the ordinary affairs of daily life) take it for granted that their thoughts, beliefs, and desires are the reliable antecedents of actions that are intelligibly related to these very antecedents. The metaphysical conundrum arises when we ask how these seemingly nonphysical, nonspatial antecedents somehow operate in the realm of physical matter and energy. There is a kindred way of addressing this question. We can ask just how it is that uttered noises, reaching our ears in the form of intelligible words and phrases, persuade us to take decisive actions sometimes extending over periods of years and decades. That is, we can ask just how it is that *rhetoric* somehow *moves us*. I choose rhetoric rather than, say, music or painting, for these arts might (arguably) be said to have their effects at levels of nondeliberative function. Rhetoric, on the other hand, requires the form of argument and perceived elements of logical structure and organization. To reserve some pages, therefore, on how rhetoric yields actions is to provide a realistic framework within which to consider how the "Cartesian" problem might be profitably understood.

Long before Aristotle gave the subject its disciplinary form, rhetoric was understood by the Sophists and by Socrates (and later by Anscombe) as an applied *psychology* whose efficacy depended entirely on the adequacy and soundness of the foundational subject. But among the ancients it was Aristotle who most thoroughly examined the sources of rhetorical influence and refined the subject, as he did so many others, into a systematic body of knowledge and theory: a technical study, as he says, whose subject matter is the modes of persuasion.[3]

I will focus here on books 1 and 2 of the *Rhetoric*, where the modes of persuasion are associated with the spoken word, and I will examine just two of the three factors identified by Aristotle as supplying rhetoric with its persuasive force. The three factors cited by Aristotle are the character of the speaker, the mental or affective states into which auditors might be moved, and the actual logical or evidentiary resources contained in the rhetorical performance.[4] These establish for Aristotle the conditions of success needed by the rhetorician. I shall refer to them as the conditions of credibility, receptivity, and proof, though I will devote most of what follows to the first two of these. Before turning to the first two, I should comment briefly on the third.

The role of the logical structure of a rhetorical argument has been examined closely in a most informative way by Eugene Garver.[5] He recognizes that Aristotle (unsurprisingly) locates the essential nature of rhetorical undertakings in the ends sought rather than in the purely formal properties. There are, indeed, distinctions to be made between and among the types of rhetoric (*vide infra*), but what makes any of them "rhetoric" is the purpose of the endeavor. Nonetheless, for Aristotle form and function are tethered in the natural world. How are they linked in rhetoric? Just how is it that logical structure and reasoning persuade? To this significant question, Garver offers replies suggested by Hume's famous discussion of *Why Utility Pleases*.[6] Garver presses on past the obvious and reveals a set of nested difficulties when the analysis turns to rhetoric's aims and reasoning's formal structure. For Aristotle, there is no "reasoning" apart from actual reasonings; i.e., apart from deliberating, judging, praising, blaming, etc. But these activities are distinct from the formal structure, the actual logic of the arguments set forth in the service of these aims. How is the seeming disparity collapsed, if at all? The most plausible answer given by Garver—and by Aristotle—is that "the character of the speaker *is* what is revealed in the speech. . . . Reasoning persuades because it is evidence of φρωονησις and character."[7] No more needs to be said regarding this third factor, then, for we see that it is incorporated into the first. If the perceived character of the speaker is a significant factor in determining the success of rhetorical performances, then the rational structure of the argument derives its power by way of illuminating the character and practical reasonableness of the speaker.

Clearly, although the three factors are distinguishable, they are nearly invariably interrelated and, to some extent, interdependent in practice. Consider only how the audience's mental or emotional or moral state might incline them to misjudge the character of the speaker; or how the evidence and logical rigor of an argument might alter the emotional tone of an assembly. Yet again, the command of relevant evidence and the precision and judiciousness with which it is deployed by the speaker may cause auditors to impute an exceptional character to him. Against the opinions prevailing among writers on rhetoric in his own time, Aristotle takes the *character* of the speaker to be the primary source of rhetoric's power of persuasion.

In observing this, however, Aristotle finds something both odd and irksome. After all, under ideal conditions those perceived to be of upstanding character, those who are "good men," should enjoy neither more nor less credibility than what their words warrant. Persuasion, he says, "should be achieved by what the speaker says, not by what people think of his character before he begins to speak."[8] Aristotle offers illustratively the deliberations of the judges of the Areopagus and other courts in what he calls well-governed states. In such settings disputants are simply not allowed to talk about nonessential matters. "This is sound law," he says, for it is wrong "to pervert the judge by moving him to anger or envy or pity—one might as well warp a carpenter's rule before using it."[9] There is a theoretical issue here, as well as a seeming paradox tied at once to considerations of a purely formal nature and at the same time to others that are personal, political, and ethical. One should be persuaded by the quality of an argument and should be otherwise immune to the reputed or established honor of the speaker. But the latter proves to be the major source of rhetorical influence.

In light of this, one might be confused as to how Aristotle meant the subject of rhetoric to be understood. Is it an effective device by which to achieve ends otherwise blocked by purely judicious reasoning? Is it a way to move those whose passions are easily stirred? Is Aristotle to be counted among cynical and even Machiavellian experts informing those who would aspire to high office just how to manipulate an audience? Indeed, there is much in the essay that presents itself as just such a "how-to" manual. There are, however, constant reminders in the same pages of the qualities of the virtuous person and the emotions and dispositions of the vicious. Examining these passages, one is inclined to regard the *Rhetoric* as a means of knowing oneself, knowing one's vulnerabilities in the face of masters of persuasion. Taking both aspects of the work at once, one does find the elements of paradox, for the same essay seems to recommend what it condemns.

How is the paradox, if there is one, to be unraveled and settled? To begin, the *Rhetoric* is concerned with a special kind of art that must guide us, as Aristotle says, where we have no other art or systematic knowledge on which to rely.[10] As persuasion itself enters in nearly every aspect of social life, rhetoric is omnipresent. Thus a taxonomy is needed just to organize the subject, but this very taxonomy depends on that most general of the perspectives to be found in Aristotle's nonlogical treatises; viz., the *ethological* perspective. The way to go about defining and identifying the principle forms of rhetoric is to examine its various aims and occurrences where it is explicitly practiced. To know what it is intended for and to know the causal modalities on which it depends is finally to understand what kind of art or science it is. This part of the inquiry is essentially value neutral. One can examine the purposes of a trireme and the mechanical and dynamic principles by which it becomes seaworthy without approving of such acts of warfare or piracy as might be perpetrated by the crew. At this point, then, we are best served by reviewing the taxonomy.

Aristotle partitions rhetoric into three branches. With his characteristic passion for categorization, he distinguishes between and among what he calls the deliberative, the forensic, and the epideictic forms of rhetoric, each operating within a definable temporal sphere. Deliberative rhetoric urges action or forbearance and thus seeks to influence the future, chiefly by appeals to prudence and utility. Forensic rhetoric, on the other hand, is the rhetoric of prosecution and defense, the rhetoric of adjudication, and is based on events that have already taken place and which now cry out for justice. Epideictic ("showy" or "flowery") rhetoric differs from both of these. It is the rhetoric of censure or encomium, the rhetoric of the here and now, seeking to honor or to condemn.

Note that both the timeframes and the ends sought by these three forms of rhetoric differ. We often honor those who throw caution or prudence to the wind and engage in heroic actions that later warrant great praise even by those who first judged the effort to be futile or foolish. Similarly, one might win at litigation on a minor technicality but thereby secure justice in a manner that is worthy of censure. These specific matters aside, Aristotle encourages us to consider in the most general terms the basis on which we should be urged to do anything at all or on which urgings are likely to be heeded. Just what is the most general objective of exhortations? Unsurprisingly, this turns out to be happiness and its constituents.[11] Whether the rhetoric is deliberative, forensic, or epideictic, it must finally make contact with commonly held desiderata, among which Aristotle includes good birth, good friends, wealth, good children, health, beauty, honor, and a happy old age, adding a measure of good fortune or luck.

No reasonable person would urge others to neglect or undervalue these, nor would an audience be easily moved to action were these and kindred aims not at stake. But once "happiness" or the good life is taken to be the end or target or goal of an activity, and rhetoric is no exception, the nature of happiness itself must be clarified, as well as the nature of that beneficiary or seeker who would find happiness in one or another form of life. Whether we examine individual persons or entire governments, says Aristotle, their qualities "are revealed in their acts of choice, and these are determined by the end that inspires them."[12] When rhetoric is able to move them to choose, their choices reveal not only the power of rhetoric but also something fundamental about those moved by it. So the subject of rhetoric turns out to be more than an examination of the modes of persuasion. It is also, and perhaps more importantly, a means by which to evaluate the qualities, chiefly moral, both of those who exert and of those who respond to rhetorical force. It becomes part of that enlarged and *characterological* psychology found throughout Aristotle's political and ethical compositions.

If the character of the speaker is often singularly effective in persuading others, it is of importance for the rhetorician to know how character and honor and nobility are comprehended by others. This is especially so where the rhetoric is epideictic. Just what is that excellence of character (αρετη) that so fully establishes an allegiance on the part of those who behold it? The general view, says Aristotle, is that it is what finds one *providing and preserving good things.*[13] Persons of normal judgment and perception regard as excellent those whose efforts bring about and nurture that which is good. However, for the multitude to reach this judgment it is necessary that they regard these consequences as intended. Adventitious good offers no basis for praise or honor. Thus, when the character of the rhetorician is sufficient to win others to his cause it must be a character reflecting or thought to be composed of certain dispositions. Aristotle lists them: justice, courage, temperance, magnificence, magnanimity, liberality, gentleness, prudence, wisdom—in a word, just the virtues extolled in both the *Nichomachaean* and *Eudemian* ethics.

To be sure, it is easier for the rhetorician to move an audience by ascribing such virtues to himself or to the audience itself. As Socrates noted, it isn't difficult praising Athenians before a group of Athenians![14] Lest Aristotle's observation here be taken in a cynical vein, however, note that only those won over to such virtues will be moved by those allegedly possessing them. The fool is not attracted to wisdom, nor is the small-minded to magnanimity. Again, the observation establishes one of the grounds on which to assess the moral qualities of both speaker and audience. Those who are moved by guile and pretence, or by expectations of what the virtuous judge to be mean and sordid, present the objective observer with all the evidence needed to reach a sound judgment as to the worth of the collective. To make such judgments, however, one must be have access to more than the actions themselves or even their consequences. Good persons can bring about the undesirable inadvertently. The same is true of foolishness and small-mindedness or other vices great and small, which, like their virtuous opposites, are grounded less in the act than in the dispositions that give rise to it. As Aristotle notes, "the worse of two acts of wrong done to others is that which is prompted by the worse disposition. Hence, the most trifling acts may be the worst ones."[15]

The mission of rhetoric in such cases is clear, at least within the forensic domain, where punishments would otherwise be mechanically tied to physical actions and material injuries. To fail to examine the underlying dispositions—the character of the actor—is to treat such behavior as merely a bare fact with no meaning beyond itself. Understood this way, however, rhetoric itself has a certain reflexive property, in that it seeks to bring about otherwise refractory actions and can do so only by appealing to certain dispositions. The rhetorician finds the principle to which the relevant dispositions are attached and in this way brings about a desired response on the part of those sensitive to the appeal.

To refer to what moves others is to refer to some cause of activity. Accordingly, a theory of rhetoric is also (or is based upon) a theory of action. Aristotle reduces the etiology of action to seven items, which he denominates chance, nature, compulsion, habit, reasoning, anger, and appetite.[16] Other candidate causes of action prove to be mere correlates. Thus, young men might act out of strong passions, but it is the anger or the appetite and not their youth that *causes* the action. What chance brings about has only an indeterminate cause and must be without purpose. Nature yields utterly determined and uniform outcomes beyond the reach of rhetoric, etc. If, then, the rhetorician is to induce action—and granting that he cannot do so by relying on chance or nature or compulsion or habit—he must achieve his aims by appealing to the rational, affective, or appetitive in his audience. Again, what is promised is what is or is taken to be a source of happiness; what is promised is what is pleasant, or what will end suffering or grief, or what will secure that which is commonly desired either for its own sake or for its usefulness.

The actions that rhetoric would elicit, however, have consequences not only for the actors but for those toward whom the actions are directed. As a result of this, the rhetorical context is one in which considerations of justice often arise even when the context is not forensic. Justice for Aristotle is not univocal. It refers both to statutory requirements and to abiding principles of equity, to what is local and expressible in writing and to what requires deeper and more intuitive sources of judiciousness, to understandings that might be implicit in the legislative intent but might also reach cases never envisaged by the lawgiver.

In this the paradox dissolves further, for we see that forensic rhetoric, far from being a manipulative ploy operating extralegally, is in fact a much needed device for enlarging the juridical context to include principles of equity. What the assembly must be made to understand is what an act betokens about the actor himself. The punishment, if there is any, for injuries caused by accident or inadvertence is to be far lighter than that arising from turpitude and malignant intent. Thus, in drawing attention to what an evil person really was seeking to achieve—even where the offense itself is minor and its immediate consequences trivial—the rhetorician leads his auditors, by argument and empathy, toward conclusions faithful not only to the facts but to their significance. Thus, even in instances in which a man has acted rashly or intemperately and should therefore receive punishments or rebukes, it is equity, as developed by the rhetorician, that would "ask not what a man is now but what he has always or for the most part been. It bids us to remember benefits rather than injuries, and benefits received rather than benefits conferred."[17]

This same rationale operates in other rhetorical contexts as well for, in assigning praise or blame or in urging others on toward a nobler or happier state, one must identify what may only be implicit or even lacking in the aspirations and deliberations of one's audience. In these more general contexts, the rhetorical devices will be those Aristotle called "nontechnical": devices not formally logical, not closely tied to evidence, eyewitness accounts, or written or uttered oaths and contracts. In these same general settings the rhetorician must be informed as to how such nonrational (though not *irrational*) factors as anger and appetite move us. Accordingly and illustratively, I should say something at this point about Aristotle's understanding of the relationship between the emotions and motives that may serve as the efficient causes of behavior and the moral weight of the actions thus generated. One of Aristotle's chief topics in this connection is anger. To understand how he treats of it in the *Rhetoric* is to understand his general position on the emotions.

To begin, Aristotle is surprisingly tolerant and utilitarian, even somewhat positive, in describing affective states. In other works in which the emotions are considered under a moral light, he is careful to distinguish between that which is properly accorded moral weight and that which may be said to arise more or less naturally. The emotions and appetites are endemic to animal life and are utterly natural. They have the instrumental function of impelling actions typically associated with self-preservation and propagation. The question as to whether anger or affection or shame or fear is good or bad, right or wrong, praiseworthy or blameworthy, is unanswerable, therefore, for the emotions *qua* emotions can have no such properties. It is only when their objects or sources are identified that a moral appraisal becomes possible. To know, for example, that Smith is experiencing the emotion of love, and to know no more, is a morally neutral fact with no more significance than an itch or toothache until we discover just what or whom Smith loves. Only then is it possible to judge what or what kind of person Smith is disposed to love and therefore what kind of person Smith is.

The moral quality that can be assigned to affective states is determined by an analysis of the agent's dispositions (‘εζεις). Properly recast, therefore, our moral ascriptions will be of the form: "Smith has a good (or bad) disposition to anger (or love, or shame, or fear, etc.)." The virtuous man is angered by villainy; the great-spirited man has contempt for small-mindedness, etc. The dialectical relationship is again apparent in Aristotle's briefer discussions of pity and indignation. Pity is the pain we feel when an evil befalls one who does not deserve it. Indignation is the pain we feel when good fortune befalls another who does not deserve it. In both cases, we may be said to have a good ‘εζσις for the felt emotion. Both of these feelings, says Aristotle, "are associated with good character."[18] Tranquility in the face of injustice, to paraphrase a political slogan of not too many years ago, is no virtue!

The emotion of anger finds a special place in Aristotle's *Rhetoric* because it is most closely associated with the gaining and preserving of honor, with the avenging of injuries, and with the burdens faced by those who would have prudence prevail in times of unrest. Anger thus figures centrally in the three main branches of rhetoric: the deliberative, the forensic, and the epideictic. Perhaps more than any other emotion, it is anger that the rhetorician must understand most fully.

Anger, says Aristotle, is a species of desire: the desire for revenge on the part of one who has been slighted by those who have no cause to do so. Thus, it is always directed toward a specific person or group.[19] One may be slighted in different ways: by contempt, by spite, and by insolence. The rhetorician who would calm his audience must work against the causes of their anger; either by exhausting their anger on another target, or by promising justice in the end, or by substituting some other form of gratification. Once more, however, success will depend not only on the skills of the rhetorician but on the nature of those to whom he makes his appeals—their essential character. Nothing good can be said of those whose rightful indignation is eliminated by humor or distracted by scapegoats.

In his ethical treatises Aristotle provides painstaking examinations of the specific virtues and vices and makes clear how these must be examined with great particularity in assessing individual persons. But the rhetorician, by the very necessities of the case, must deal with collectives and must therefore possess knowledge based on generalities about human nature. He must have a typology that is serviceable in persuading identifiable collectives.[20] It is in his *Rhetoric*, then, that Aristotle begins to develop a comparative psychology of "types" at the most general level. One "type" of personality is the result—all other relevant factors being equal—of age. The young, the old, and those in their prime generally have distinctly different dispositions. The young are passionate and easily offended, their lives spent in expectation rather than in memory. Not often having been humbled, they are given to lofty notions and are inclined to be contemptuous of merely useful ones. Those past their prime are the opposite, neither loving warmly nor hating bitterly. They are sure of nothing and underdo everything. "They are small-minded," he says, "because they have been humbled by life; their desires are set upon nothing more exalted or unusual than what will help to keep them alive."[21] Those in the prime (physically, those about thirty-five and mentally those about forty-nine) are found between these extremes—their lives guided "not by the sole consideration either of what is noble or what is useful, but by both; neither by parsimony nor prodigality, but by what is fit and proper."[22]

Clearly, the well-ordered state is architectonic for the well-ordered life (and vice versa), which tends to be the life lived by those in their prime. It conducts its affairs not so much "moderately" but mid-way between excess and deficiency. Were all citizens of the polis paragons of virtue, there would be no need for rhetoric. A rational analysis of the facts would combine with the virtuous dispositions to yield exemplary conduct. Alas, this is an unrealized ideal. It is the power of rhetoric, rightly deployed, that can overcome the limitations of youth or senility, of rashness or cowardice, of tendencies to vice.

There is no clear and effective alternative to these powers except under any but a totalitarian regime. Where democratic principles are valued, where justice itself is valued, where meritocratic principles are installed, there one will find a place for rhetoric. Neither wealth nor luck nor a good pedigree will serve in its place. Luck is fickle and the wealthy soon come to regard wealth as the only standard of value. As for breeding, Aristotle deserves to be quoted at length: "Being well-born, which means coming of a fine stock, must be distinguished from nobility, which means being true to the family nature—a quality not usually found in the well-born, most of whom are poor creatures. In the generations of men, as in the fruits of the earth, there is a varying yield."[23]

Where Socrates would breed the Guardians and rely on eugenic programs to populate the ideal state, Aristotle, ever the realist, considers various ways of making the best of an admittedly dangerous and unpredictable situation. He is satisfied that he understands the nature of virtue and even the conditions that instill it in those (few?) who are receptive. But this leaves unattended that great mass of humanity on whom the polis must finally depend for its very survival. If guidance is to be effective, the teacher must know his students' strengths and limitations, aversions and temptations. In all, whether or not one is to be *moved* to action is to be understood in terms of determinants and *self*-determinants bound up with character, with context, and with nothing short of the influences and the initiatives of a lifetime. This is the position reached by an ancient philosopher otherwise strikingly naturalistic and biological in his approach to creature-level modes of movement, desire, and feeling. It is, however, a position that realistically reaches those complex social and cultural levels at which the factors impelling action are not credibly reduced to creature-level modes of behavior.

There is in this part of Aristotle's philosophy something of a parity between the skilled rhetorician and that rational being who would lay out a reasonable course of action for the "natural slave" of the *Politics*, whose mere mention heaps such mountains of scorn on Aristotle. The natural slave is one who, though lacking full powers of rationality, is able to follow the rational dictates of others. And the masses? They, too, lack the full power of the man of virtue but can be moved toward the sorts of actions habitually performed by the virtuous. Even for this much to be possible, however, there must be something in them that inclines them in this direction. The skill of the rhetorician is in tapping this vein of inclination and in setting up a positive disposition for its full expression. In this way an age becomes known and might even come to know itself, for we are known, says Aristotle, by our choices and by what moves us. With the respect due to those who look to science for explanations here, I would say that the only level at which to comprehend and explain the motivating power of rhetoric is, alas, the *rhetorical* level, which must include estimations of the character and virtue, the aims and values, the culture and the judgment of the participants, including the speakers themselves. How rhetoric "moves" us is not a question properly seeking biomechanical explications, which, in the circumstance, would surely seem droll.

Let me return now to Descartes and "Cartesianism." The challenge to Descartes requires his dualistic ontology to account for how "mental" events could "cause" muscles to move. Stated this way, we see that the challenge is no more than an attempt to absorb deliberation, desire, and motivation into a mechanistic framework. The attempt itself calls for an explanation, for it is not at all obvious that questions of this sort would ever arise outside the domain of philosophical speculation. It might be asked why, at the level of daily life, persons would be inclined to think in *causal* terms at all when considering relations between mental and bodily events. With commendable patience, Descartes attributed this to the fact that his critics (not to mention ordinary persons) had never had the experience of mental life apart from their bodily states. As he says in his second set of Replies:

> If as may well be the case, they take the view that the formation of thoughts is due to the combined activity of parts of the brain, they should realize that this view is not based on any positive argument, but has simply arisen from the fact that, in the first place, they have never had the experience of being without a body and that, in the second place, they have frequently been obstructed by the body in their operations. It is just as if someone had had his legs permanently shackled from infancy; he would think the shackles were part of his body and that he needed them for walking.[24]

Descartes' analysis here is not unlike the explanation one would offer of *phantom limb*. Without exception, every mental event has occurred to us in our *embodied* state, such that the *mental* association between the two finds no exceptions. In *phantom limb* the pain is felt in a limb no longer attached to the body, but sensations of this kind in all of previous life to date have always been associated with the pained limb *of that body*. By a kind of fixed memory template, it is extremely difficult to reconcile the pain of *phantom limb* to the fact that there is no limb. Similarly, it is extremely difficult during an embodied lifetime to comprehend the very possibility of *disembodied thought*. One gambit available as a defense, even if one of evasion, is by way of Hume's challenge to the very notion of causes. On

Hume's account, the sole evidence that may ever be adduced in support of the claim that X is the cause of Y is that the two have been "constantly conjoined" in experience. Causation is thus reduced to no more than empirical regularities and the "Cartesian" problem is thereby defused. Among the most stable of experienced regularities are those in which events of a mental sort precede events of a bodily sort, and that's the end of it. I will return to this later in the chapter and offer a fuller account of Hume's analysis, suggesting that it might be just the one Descartes could accept as an explanation not of causation but of our *concept* of causation.

Convenient though this gambit is, it was not one that Descartes could adopt even if Hume's account had been available to him. Descartes was a realist about causation and certainly understood it to have firmer metaphysical grounding than common experience. Indeed, the philosopher who worried so much about an evil demon setting out to deceive him by rendering all experience merely illusory would be among the first to accept the illusory nature of any causal chain having no more than experiential links. Accordingly, in his replies to critics and in his letters to Princess Elizabeth, Descartes readily acknowledged the severity of the task associated with explaining such causal relationships as obtain between a *res cogitans* and a *res extensa*. In the well-known letter dated May 21, 1643, he admits that his efforts to distinguish between body and soul were at the cost of neglecting the very question Princess Elizabeth has raised; viz., how, on his account, one should "conceive the union of the soul and the body and how the soul has the power to move the body."[25]

To speak of "the power to move the body" is, in the most fundamental sense, to speak of *motives*, or *what moves us*, whether the causes are in some complex fashion rhetorical or in a seemingly more elementary way physiological. Descartes' critics were not at a loss to see the problems faced by a theory that requires a massless unextended "something" to move a muscle—or a person. To answer them, and in a manner that would be clear to a highly intelligent and ostensibly neutral party such as Princess Elizabeth, Descartes begins with a summary of the very grounding of our knowledge claims. He addresses the point in his letter to the princess. If he is ever properly charged with holding a theory of innate ideas, it is at this point, where he speaks of "certain primitive notions which are as it were models on which all our other knowledge is patterned."[26]

Granting that there are very few such notions, they are nonetheless the cognitive preconditions for all other knowledge. At the most general level, to have any knowledge at all is to possess notions of number and duration. If our knowledge then pertains to bodies, the foundational notion is that of extension, which "entails the notions of shape and motion." Knowledge regarding the soul, on the other hand, "includes the conceptions of the intellect and the inclinations of the will."[27] Finally, considering body and soul together, "we have only the notion of their union, on which depends our notion of the soul's power to move the body, and the body's power to act on the soul and cause sensations and passions."[28]

We might read Descartes at this point as invoking what in the next century will be a fixture in Thomas Reid's "Common Sense" philosophy. Descartes—in the spirit of the mathematician—recognizes that all systematic knowledge begins with certain core precepts or concepts that are not themselves reducible to anything more fundamental. Cartesian psychology, if we may call it this, regards as irreducible certain basic notions in the absence of which we would not be able to comprehend anything. Our awareness of our "intellect" and our "will" is not by way of empirical evidence or rational deduction. In point of fact, the absence of intellect and will would obviate the very possibility of evidence and of conclusive arguments. That we have a mental life and that we are embodied stand at the very foundation of all conceptions and actions. And, from the fact that our intentions and desires yield predictable action patterns, we develop the notion of a union between the two.

Princess Elizabeth found this rather thin as an explanation and pressed Descartes to make his position clearer. The letter of May 21 had left the matter more or less at the level of stipulation: a *foundationalist* account not unlike what is commonly found in mathematics. There are primitive notions on which all further understanding depends, primitive notions not reducible to anything more fundamental, primitive notions generative of clear and distinct ideas. These "intuitions" in place, it is then imperative to apply the fundamental notions properly, "for if we try to solve a problem by means of a notion that does not apply, we cannot help going wrong. Similarly, we go wrong if we try to explain one of these notions by another, for since they are primitive notions, each of them can only be understood by itself."[29]

This is where Descartes might have left the matter had the princess not persisted. Now addressing her reservations a month later (June 28, 1643), he apologizes for having left out so much in his earlier letter and sets about to clarify the main points.[30] He begins by rehearsing those three primitive notions or ideas that are unique in the epistemic sense: the idea of soul, the idea of body, and the idea of the union or interaction of the two. He argues for their uniqueness by declaring that the idea of "soul" can be conceived solely by way of pure intellect. He then says that "body," although comprehended by the intellect alone, is more fully comprehended when intellect is aided by imagination. Finally, the idea of mind-body interaction "can be known only obscurely by pure intellect or by intellect aided by imagination, but it can be known very clearly by the senses."[31]

The evidence favoring a sensory awareness of mind-body interaction is that ordinary persons without any philosophical orientation are fully cognizant of the power of the mind to move the body, for they have sensory data to this effect all the time. Had Descartes the benefit of contemporary reflections on these matters, there is good reason to believe he might have been especially inclined toward certain forms of *functionalism* such as that propounded by John Heil.[32] There are innumerable ways that some type and intensity of pain can be brought about in a person. To the extent that every precursor of the sensation of pain includes some measure of tissue damage, we can say that it is at the (low level) of cellular physiology that the (higher-level) property of "pain" is grounded. Not merely any (Humean) correlation or regularity qualifies here, for only a certain "causal profile" adequately accounts for "pain" in relation to myriad precursors. Pain, then, is "realized" through the operation of lower-level events, though the specific "realizing" events might well be different in different persons, not to mention different species. If Descartes' insistence that only pure intellect can comprehend "soul" is to be fairly assessed, it is important to understand what he means by *imagination*. Something is imagined when it is "*represented by a corporeal image*."[33] As "soul" cannot be thus represented, it cannot be imagined. Having neither odor nor weight nor color, etc., the soul cannot be an object of the imagination. It can be conceived, however, for (as in mathematics) all sorts of noncorporeal entities are conceivable. Indeed, "since it is by means of it that we conceive all other things it is itself more conceivable than all other things put together."[34] All conception is "about" something; in the argot of a later age, the contents of consciousness are *intentional objects*. To conceive of a blue sky is to have an idea that is not itself "blue." It is solely as entities with mind as such that we are able to represent and form conceptions of things knowingly, thereby understanding that we are in fact conceptualizing. There is no representation of a material object that *can* be understood as having this power. Thus, we cannot "imagine" mental powers, for they are not reducible to merely material properties.

In the clarifying letter to Elizabeth, Descartes goes on to say that pure intellect is also able to conceive of extended things but that our conceptions are more fully informed when supported by imagination. To put the point plainly, we acknowledge that all of Euclidean geometry can be comprehended without ever putting a mark on a piece of paper but that there is a better comprehension of the system when its axioms and theorems are represented in extended form. As for the soul's (mind's) interaction with the body, what more is needed than the incessant evidence of the senses as one's decision or desire to act is followed reliably by movement of the limbs and by (intended) effects on the external world? Thus does the ordinary percipient, without benefit of metaphysical tutors, come to regard the very union of the mental and the bodily as if there were but one entity rather than an entity in which two radically different substances were at work. He fully appreciates Elizabeth's admission that she would find it easier to conceive of the soul as extended than to think of it as an immaterial "something" nonetheless moving and being moved by the body. He assures the princess that this is a natural conclusion and even one that is useful to entertain. But after she has become comfortable with the idea that there is total unity, not a duality of mind and matter, her further consideration will then challenge this conclusion: "Once she has formed a proper conception of this and experienced it in herself, it will be easy for her to consider that the matter attributed to the thought is not thought itself, and that the extension of the matter is of different nature from the extension of the thought."[35] The materially extended thing is in a space that excludes all other extended things. Not so with the thought about it.

This is where Descartes left things circa 1650, some three hundred and fifty years before Tim Crane composed a clear summary of the issue and considered the progress that might be claimed in settling it.[36] Here is the puzzle identified by Crane in assessing physicalist theories of mental causation:

> On the one hand, the original motivation for physicalism was the need to explain the place of mental causation in the physical world. On the other hand, physicalists have recently come to see the explanation of mental causation as one of their major problems. But how can this be? How can it be that physicalist theories still have a problem explaining something which their physicalism was intended to explain in the first place? If physicalism is meant to be an explanation of mental causation, then why should it still face the problem of mental causation?[37]

Alas, the "problem" of mental causation. Tyler Burge has written on this with great discernment and it is with full respect that the following passage is cited and then critically assessed:

> I will assume . . . at least that mental states and events would not occur if some "underlying" physical states and events did not occur. There are no gaps in these physical chains of events. So, for example, there are underlying, gapless neural processes that are instances of laws of neurophysiology; and the mental events would not occur if some such processes did not occur. I have no serious doubts about this view.[38]

I do have serious doubts about this view, for I have serious doubts about the aptness of causal modes of explanation in accounting for events arising from mental states. Burge notes that relationship between physical and mental events "is thoroughly unclear"[39] and concludes that there may be something amiss in current models of physical causation. It is at least courageous to equivocate here. Perhaps, beyond courage, there may be room for a certain rashness and simply affirm the uniqueness of the mental. After all, when centuries of speculation and decades of painstaking analysis have yielded no evidence whatever to favor the conclusion that mental life is in any way physical in its character, it is timely to adopt, if only provisionally, the position that the reigning "causal" model of explanation in physics has nothing much to offer here.

Physicalism now has the same difficulties as those adumbrated by Descartes in his letters and his replies to critics. One difficulty leads to question-begging assumptions that redeem physicalism by supposing its truth. Illustrative of this gambit is the assumption David Lewis makes about the explanatory adequacy of physics. The argument, sifted to its essence, is that behavior is a physical event and all physical events are physically caused. There may be nonphysical items, but these cannot be part of any causal account of physical outcomes. Thus, if "mental" events have causal power, they must be physical.[40] However, as the issue of "mental causation" derives its force from the seeming limits it imposes on the reach of physicalism, it is scarcely helpful to relax the tension simply by declaring that physics is adequate to the task of explaining all physical events! It is worth raising other questions about the sense in which the behavior of consequence is physical. Surely the movement of muscles and the overall pattern of gyrations displayed by musculoskeletal systems are physical, but in the same sense that a dead body dropped from a height—as a physical body—will conform to the laws of classical physics. This is as much the case with a body thus dropped as with a suicidal leap from the same height. If all that requires explaining is the rate of descent, then no more than Galileo's formulations are necessary. But the "behavior" arising from suicidal motives is obviously not merely a physical event.

Surely not a disciple of Descartes, Tim Crane concludes his instructive essay thus:

> There is no good reason for saying that . . . mental phenomena are "constituted by" or "realized by" physical phenomena . . . it's only insofar as mental states have effects *in the very same sense that physical states have effects* that we need to think of them as physical states. . . . [But] orthodox physicalism denies that mental states have effects "in the very same sense" that physical states have effects. For it now says that there are mental phenomena which are causes in their own way; and there are physical phenomena which are causes in a different way. Whether or not such distinctions between different kinds of causation are ultimately tenable, it seems plain that little is added by saying that these mental phenomena are "ultimately physical." And it seems an empty terminological decision to call the resulting position "physicalism"—except perhaps to put on the record one's differences with Descartes.[41]

Was Descartes not making the same point when he said, "we have hitherto confounded the notion of the soul's power to act on the body with the power one body has to act on another"?[42] The confounding arises from assimilating all mental powers into the framework of physical powers but then offering no plausible explanation for the manner in which what is patently *mental* results in the given course of action. Persons are moved by desires and beliefs, by having good reasons to act and reasonable strategies for securing various desiderata.

In addition to actions coming about as a result of motivation, persons in their ordinary lives explain their initiatives in terms of desires and beliefs. Desiring to attend to her thirst, Mary moves toward the fridge in the belief that a pitcher of cold cranberry juice will be found there. Asked to account for Mary's movements, one may take the request as a search for causes, but this would raise additional questions regarding the mode of causality assumed in the resulting account. That she was moved by desire and belief seems clear enough until we confront the Princess Elizabeth conundrum of a mental state overcoming the inertial constants of that embodied mass denominated "Mary." Apart from the dispute between physicalists and dualists, there is the comparably fundamental question of just how causation itself is to be understood and how it is in any defensible sense known. I return once more to Hume, now at greater length and with the intention of showing how the "Cartesian" problem of metal causation might be recast. I've noted that Descartes himself would not adopt the Humean reduction of causality to regularity, but he would (and did) accept what Hume took to be the grounding of our *concept* of causality.

Following Locke, Hume identified the principal modes of knowing as intuitive, demonstrative, and perceptual. On this scheme, it may be asked whether we know the cause of things intuitively, through a demonstrative argument, or solely by experience. The importance of the question arises from the very nature of what we take to be our knowledge of the world; viz., our identification of causal dependencies by which we can manipulate and predict the behavior of objects in the natural world. The Pythagorean theorem expresses what Hume refers to as "relations of ideas"—the relationship between the square of the hypotenuse and the sum of the squared sides. As this is part of the larger formal argument of Euclidean geometry, its validity is independent of anything in the actual world of matter and thus is known through "the mere operation of thought," as Hume says. Distinct from such relations among ideas, however, are *matters of fact*,

> which are the second objects of human reason, are not ascertained in the same manner; nor is our evidence of their truth, however great, of a like nature with the foregoing. The contrary of every matter of fact is still possible; because it can never imply a contradiction, and is conceived by the mind with the same facility and distinctness, as if ever so conformable to reality. That the sun will not rise tomorrow is no less intelligible a proposition, and implies no more contradiction than the affirmation, that it will rise. We should in vain, therefore, attempt to demonstrate its falsehood. Were it demonstratively false, it would imply a contradiction, and could never be distinctly conceived by the mind.[43]

Hume goes on to say that all factual reasonings are based on causal relations, permitting us to extend our understanding beyond the immediate evidence of the senses or our recollection. We make inferences regarding how things will be, for we assume, on the basis of regular experiences, that events and objects will behave as they have been observed to behave, and this because:

> It is constantly supposed that there is a connexion between the present fact and that which is inferred from it. Were there

> nothing to bind them together, the inference would be entirely precarious. . . . I shall venture to affirm, as a general proposition, which admits of no exception, that the knowledge of this relation is not, in any instance, attained by reasonings a priori; but arises entirely from experience, when we find that any particular objects are constantly conjoined with each other. Let an object be presented to a man of ever so strong natural reason and abilities; if that object be entirely new to him, he will not be able, by the most accurate examination of its sensible qualities, to discover any of its causes or effects. . . . When I see, for instance, a billiard-ball moving in a straight line towards another; even suppose motion in the second ball should by accident be suggested to me, as the result of their contact or impulse; may I not conceive, that a hundred different events might as well follow from that cause? May not both these balls remain at absolute rest? May not the first ball return in a straight line, or leap off from the second in any line or direction? All these suppositions are consistent and conceivable. Why then should we give the preference to one, which is no more consistent or conceivable than the rest? All our reasonings a priori will never be able to show us any foundation for this preference. In a word, then, every effect is a distinct event from its cause. It could not, therefore, be discovered in the cause, and the first invention or conception of it, a priori, must be entirely arbitrary.[44]

Here, then, is the famous "regularity" theory of causation, consistently empirical, utterly skeptical toward rationalistic attempts to comprehend factual states of affairs a priori. The same regularity theory may then be turned inward, allowing us to recognize ourselves as having some sort of power to bring about our own actions through what we call our will or volition. We cannot have any factual knowledge of the ultimate grounding of this power, but, through experience, we come to know the effects thus brought about. Descartes' dilemma is acknowledged by Hume, with Stoic resignation:

> Is there any principle in all nature more mysterious than the union of soul with body; by which a supposed spiritual substance acquires such an influence over a material one, that the most refined thought is able to actuate the grossest matter? Were we empowered, by a secret wish, to remove mountains, or control the planets in their orbit; this extensive authority would not be more extraordinary, nor more beyond our comprehension.[45]

We are not to expect to discover how all this comes about, only the conditions under which actions of a given kind follow mental states of a given kind. With respect to these actions, Hume says: "That their motion follows the command of the will is a matter of common experience, like other natural events: But the power or energy by which this is effected, like that in other natural events, is unknown and inconceivable."[46] Note, then, that on Hume's account, our ignorance of the power or energy by which one set of events reliably leads to another is spread across all such natural occurrences, both those originating in our volitions and those occurring in the external physical world. "Mental" causation is thus no more or less problematic than physical causation. Newton's laws express the rule by which gravity "causes" objects to fall toward the center of the earth. Just *how* this happens, just *what* the causal mechanism of gravity is, is not revealed by Newton's laws. The bearing this has on the alleged problem faced by "Cartesianism" is direct: True, Descartes cannot explain just how a mental event causes a bodily effect or vice versa, but nor can a physicist explain just how the moon's gravitational field causes the tides. It does, and that it does is established (ultimately in human terms) through observed constant conjunctions.

In section 5, part 1 of his *Enquiry*, Hume attempts to explain the basis on which causal inferences are made in the first place. He conjures a person new to the world but fully possessed of rational power. The world now regales the senses with clutter. This visitor has no means by which to locate the ultimate source of power by which one set of events is somehow generative of another. But experiences, reliable in their content, develop within this innocent person; this is what Hume calls the *habit* of inferring causes from such reliable sequences of antecedents and consequences. The visitor

> has not, by all his experience, acquired any idea or knowledge of the secret power by which the one object produces the other; nor is it, by any process of reasoning, that he is engaged to draw this inference. But still he finds himself determined to draw it: And though he should be convinced that his understanding has no part in the operation, he would nevertheless continue in the same course of thinking. There is some other principle which determines him to form such a conclusion. . . . This principle is Custom or Habit. . . . By employing that word, we pretend not to have given the ultimate reason of such a propensity. We only point out a principle of human nature, which is universally acknowledged, and which is well known by its effects. Perhaps we can push our enquiries no farther.[47]

In the spirit of a *pax philosophica*, one might suggest that Hume's position by way of custom and habit is not that different from those intuitive notions Descartes cites in his letters to Elizabeth. We are able to form notions of material things, notions of mental states, and notions of the interactions between these. Descartes surely would not scruple over the term "habit," for Hume himself treats this as the result of some underlying "something" by which custom produces causal inferences. Hume, in this same treatise, reflects on the biology of it all; he speaks of nerves and even animal spirits but quickly retreats from mechanistic speculations and finds safe haven in the understanding of ordinary persons.

At both the metaphysical and the psychological levels, the Humean approach, for all its seeming common sense, is less than convincing. Beyond the physics laboratory, "conjunctions" are seldom "constant." The real world is anything but a constellation of regularities. Things happen, we say, for the most part, or generally—Aristotle's ωσ επι το πυλι—and with insufficiently connected bonds to generate Hume's habitual inferences. J. L. Mackie would rescue the thesis by way of a probabilistic thesis: though experience is rife with irregularities, certain Humean conjunctions render our predictions more probable and the nonconjunctions less probable.[48] Ruled out by Hume and Mackie[49] is the prospect of backward causation, but then this seems to allow the very a priori constraints that empiricists are at pains to reject.

Apart from this, there is the incisive criticism mounted by Hume's contemporary Thomas Reid.[50] Though there is overlap in their positions, there are notable and fundamental differences, with Reid often misleadingly understood as an "intuitionist."[51] Reid grants that it is through experience and by way of Hume's "associations" that one connects freezing temperatures to the formation of ice. If low temperatures have this effect now and have had this same effect reliably in the past, we do not doubt but that under such conditions water will freeze in the future.

> That this is a truth which all men believe as soon as they understand it, I readily admit; but the question is, Whence does its evidence arise? Not from comparing the ideas, surely. For when I compare the idea of cold with that of water hardened into a transparent solid body, I can perceive no connection between them. . . . True, experience informs us that they have been conjoined in time *past*: but no man ever had any experience of what is *future*: and this is the very question to be resolved. How do we come to believe that the *future* will be like the *past*?[52]

The mere formation of some sort of mental connection between and comparison of two different ideas—that of coldness and that of the hardness and transparency of ice—is explicable in terms of experience. But the belief that the future is under some constraint such that past regularities will recur is, itself, not formed by perceptual experiences. No number of previous connections could by themselves generate a belief of this nature. Such a belief is "an original principle of the mind" absent which we could learn nothing from experience. We would form no inferences and our memories would count for nothing.

It was Reid's position that Hume got causality backward. The mere recurrence of event pairs could not give rise to the notion of a *causal* relationship, except in a creature possessed of *active powers* itself. That we have such powers allows not only action but the withholding of action, the doing and the forbearing. Were we unaware of such power, there would be no coherent basis for our undertakings, no basis on which to hold ourselves or others responsible, no basis on which even to comprehend the behavior of others. Indeed, it is owing to the recognition of our own active power that we impute comparable power to others when they are seen doing the sorts of things we do. "A little girl ascribes to her doll, the passions and sentiments she feels in herself. Even brutes seem to have something of this nature. . . . All our volitions and efforts to act, all our deliberations, our purposes and promises, imply a belief of active power in ourselves."[53]

We impute to regularities in the external world a causal connection modeled on the connection between our own powers and the effects of their deployment. Hume, according to Reid, correctly located the question begging by Hobbes, Clarke, and Locke in their attempts to show that all things that begin to exist *necessarily* have causal antecedents. Hume's mistake, however, was in concluding that the laws of association provide the grounding for such a belief. Reid's criticism includes the observation that "causation is not an object of sense. The only experience we can have of it, is in the consciousness we have of exerting some power in ordering our thoughts and actions. But this experience is surely too narrow a foundation for a general conclusion."[54] More than this, "no reasoning is more fallacious than this, that because two things are always conjoined, therefore one must be the cause of the other. Day and night have been joined in a constant succession since the beginning of the world; but who is so foolish as to conclude from this, that day is the cause of night, or night the cause of the following day?"[55]

Again, it is a principle of mind to construct out of the data of experience a comprehension of those laws of nature by which its operations come to be predictable. For all practical purposes, the untutored mind is able to operate within the natural realm, sufficiently attuned to those regularities that match up with the dependable relationships obtaining between one's active power and one's effect on the external world. Children and those Reid's epoch referred to as savages may not rise higher in their understanding than this projection of active power into a material world actually devoid of it. Even the educated citizens of the Enlightenment might harbor notions of causality that developed science will show to be superstitions and rank ignorance. But even with the progress of scientific modes of inquiry and explanation, Reid regards it as doubtful that ultimate *causes* will be unearthed. Rather, the progress of science is measured in the discovery of those basic *laws* that describe how the fundamental causes are expressed.

Hopeful that the patient reader will not take all this to be a digression, the problem of "mental causation" might now be recognized less as a special problem than as part of a general and enduring one. Hume and Reid, opposed on so many matters, were in full agreement that Hobbes, Locke, and Clarke had not succeeded in establishing that things that come into being *must* have antecedent causes. With respect to one's own actions, Reid would have these understood as the result of the exercise of *active power*, the source of which is the actor. As he denies such power to matter as such, he would deny it even to such complex matter as, for example, brains and nerves. To say, then, that one's *brain* decides on a course of action or undertakes to accomplish something would be an abuse of language.

Motives and desires are aspects of conscious life inseparable from persons. That pathological conditions in the nervous system—or that activation of areas in the normally functioning nervous system, for that matter—alter the desires and motives of the person is unremarkable and, to the matters at issue, irrelevant. It is entirely unsurprising that, among the keen desires and powerful motives to which one's conduct is reliably responsive, the conditions and needs of the body will be the sources. There are, however, qualitatively different motives and desires: those associated with honor, ambition, self-worth, curiosity, creativity, friendship, aesthetics—the panoply of aims and objectives universally evident in human history and across widely differing cultures. Understanding motivation and desire in all these contexts, including those bearing directly on biological survival, begins with the fact of consciousness and of the actor's conscious awareness of the power to do something about a given state of affairs.

It is not obvious, in light of currently official modes of inquiry and explanation, that cognitive neuroscience has access to the business end of all this. In what most practitioners in these fields take to be the essence of scientific knowledge, the aim is the unearthing of causal laws. This, after all, is what makes sense of the otherwise peculiar attention paid to the biochemistry and functional anatomy of the central nervous system. Hand in hand with this aim is obeisance to physicalism (of one or another sort), for that is the *ism* that guarantees causal lawfulness. The overarching perspective (with a bit of room for indeterminacy at the level of subparticles) is *determinism*, no less consistent with physicalism for its probabilistic refinements. Where, one may ask, are motives and desires to be found within such a framework? And the answer, of course, is *in the brain*, for it would seem far less convincing were the answer to be *in the active life of actual persons*.

Might there not be a compromise? It was William James who coined the term *soft determinism* and who sought to avoid the further eulogizing of "freedom" by adopting the more neutral word *chance*. Against hard determinism, he offers a reality where

> the parts have a certain amount of loose play on one another, so that the laying down of one of them does not necessarily determine what the others shall be. It admits that possibilities may be in excess of actualities. . . . It says that there is a certain ultimate pluralism . . . and, so saying, it corroborates our ordinary unsophisticated view of things. To that view, actualities seem to float in a wider sea of possibilities . . . *somewhere*, indeterminism says, such possibilities exist, and form a part of truth.[56]

The ordinary unsophisticated view thus reaches for a truth at least as eagerly and perhaps more credibly than alternative views.

8

PLANS
An Epilogue

There would seem to be little evolutionary point to consciousness except as an aid, if not a necessary instrument, for planning a future. I must be clear on this. What registers in consciousness moment by moment is gone before anything can be done about it. Having some means by which to store or preserve such happenings is also pointless, unless the record is to be of use at a later time. Metaphorically speaking, and within the (arguably relevant) framework of evolutionary psychology, what counts in consciousness is a past that can be brought into it as a means by which to engage the future, often the immediate future. Consciousness thus understood is a mode of deliberation distinct from awareness. To be aware is to be in contact with the present. To be *consciously* aware is to be disposed toward all that might follow based on all that has preceded. One of the roots of "conscious" is *scio*, alerting us to the epistemic element that, when removed, leaves only awareness in its wake. It would be tempting to reduce all this to an equation: *Consciousness = awareness + knowledge*. This is haphazard. Consciousness has a focus and is therefore not in any way a passive state. It is rather a state of what might be called *directed awareness*, and in this respect would seem to have a volitional component, itself needful of knowledge, no matter how minimal. It is doubtless that the engineering required here is preestablished in some respects and forged in the kilns of experience in still other respects. *To focus is to filter.* It is to exclude all but the target. Some filtering is passive. Human vision is effective in rendering visible electromagnetic radiation falling within the range of wavelengths from about 360 to about 760 nanometers. Shorter (ultraviolet) and longer (infrared) wavelengths are not seen. So, too, with auditory sensations. Human beings can experience sounds falling in the frequency range of about twenty to about 20,000 hertz. The sensory systems thus have a passive filtering power imposed by their structural and physiological nuances and limitations. Filtering is also *selective*. We are able to maintain a conversation in a noisy room, hearing clearly what our interlocutor has to say and remaining largely oblivious to "extraneous" sound. We call the process *attention*, but in the background are surely factors best understood as volitional, motivational, and emotional. We can shift our attention at will, and we do so just in case we are moved by something more salient than the content of the ongoing conversation. By a shout of "FIRE!" for example—or even by the soft mention of our name in a distant corner of the room.

To speak of plans is to speak of activity that might be as simple and immediate as the next two steps or as complex and remote as the pianistic prowess one hopes to gain over a period of decades. The latter plan, illustrative of many that occupy large fractions of a conscious lifetime, aims at a state or condition that cannot be known until it is attained. There is no "picture" of it here and now, or tomorrow, or next year. One might in some vague sense seem to be getting closer but this very sense is less "sensory" than imaginative. *To plan is to imagine.* And to imagine is to weigh alternatives and, again, to "filter" out nonstarters or items lower on the list of things worth pursuing. The searching question arising from these ubiquitous facts of folk psychology is whether "cognitive neuroscience" is the right model of explanation, the right path to a deeper or fuller understanding of that *planning* that reflects the actively lived life, the life outside the vat.

As might be expected, Daniel Dennett speaks most clearly for the influential quarter subscribing to a functionalist perspective and the confidence it inspires in computational approaches to matters of this sort. Consider the following passage:

> Since the earliest days of cognitive science, there has been a particularly bold brand of functionalistic minimalism in contention, the idea that just as a heart is basically a pump, and could in principle be made of anything so long as it did the requisite pumping without damaging the blood, so a mind is fundamentally a control system, implemented in fact by the organic brain, but anything else that could *compute the same control functions* would serve as well. . . . If all that matters is the computation, we can ignore the brain's wiring diagram, and its chemistry, and just worry about the "software" that runs on it. In short—and now we arrive at the provocative version that has caused so much misunderstanding—in principle you could replace your wet, organic brain with a bunch of silicon chips and wires and go right on thinking (and being conscious, and so forth).[1]

The philosophers' zombies have haunted the literature on consciousness for more than a decade, their putative powers and limitations supplying grist for learned argument. The usual account has them very much like the rest of us in behavior but devoid of consciousness. Dennett's variant, fortified by the "intentional stance," possesses consciousness in whatever sense anyone or anything does, in so far as it performs (functions) as do those items uncontroversially credited with consciousness. So zombies as such fit into this picture, for they, too, may be made up of the same requisite chips by which they may well "go right on thinking (and being conscious and so forth)." Consciousness, then, pardonably the last holdover in the once thick but now paper-thin book of mysteries, is an elaborate achievement of a congeries of microcircuits, each of them in isolation devoid of mental properties but in unison the source of these very properties. Dennett draws inspiration here from the success in accounting for metabolism in such terms, though no single cell can do the job. He would be impatient, I suspect, with the reminder that "metabolism" refers to events having no physical properties not found in every cell: mass, extension, oxidation, etc. Neither the single cell nor the entire participating choir of cells offers any evidence whatever of consciousness. That turns out to be a property of mine—and yours. To ask just *how* so many cells bring this about begs the question. From all that is now known about cells, in isolation and in networks, the scientifically responsible position should be that there is no clear evidence or coherent theory warranting a physicalistic explanation. If there is any sort of warrant at all, it is one of inference from the world of already established and purely physical event-sequences. But this very inference is, again, a begging of the question at issue.

What has made zombies so interesting to philosophers? The picture here is cloudy. Some would argue that the existence of zombies must be an embarrassment to physicalists in that these seeming and utterly physical systems nonetheless have mental properties, no matter how diminished.[2] Dennett, as we have seen, argues the other way. It doesn't matter either way, however, for, if it came about that a large number of silicon chips gave rise to consciousness, the explanatory gap would remain as wide as ever. The chatter about zombies is beside the point.

What is not beside the point is whether a large number of silicon chips can ground plans and purposes of the sort that characterize actual lives. That such chips can, as it were, retain a record of such plans is doubtless; the same is true of filing cabinets and tape recorders. To ground plans (as distinct from recording them) is to imagine possibilities. Of course, possibilities can be schematized and reduced to a computational format; otherwise, Deep Blue would have been a very poor program for making chess moves. Note, however, that there is more than one sense of an outcome being possible and more than one sense of a given outcome realizing a plan. The decision to ride one's bicycle to the lecture is a conscious decision. The actual sensory-motor sequences associated with cycling (a) realize the plan but (b) unfold without conscious deliberation. In one sense, "possible" means no more than that something is allowable or permissible. A simple thermostat, composed of a bimetallic strip, will keep ambient temperature within a given range. The metals and contact points can be chosen such that a two-degree drop in temperature will find the metal bending in one direction; a two-degree increase, in the opposite direction. How the strip bends then determines whether the furnace is turned on or off. The "possibilities" here are not *imagined*. They are simply the given facts of the ambient environment, which act upon an inert element. The thermostat does not envisage possibilities nor does it plan to control ambient temperature. The conceptual drawbacks of functionalism are made evident by such devices. It is true but trite to say that two entities are functionally indistinguishable when they behave in such a way as to perform the same function. It is a theory of uncertain validity to claim that the behavior of two entities performing the same function is explicable in the same terms. (*Pace* Alan Turing.) The weather, operating over eons, surely might forge out of rock a shapely specimen with the appearance of a face; the same might be achieved by a sculptor. Only by crude metaphor would one refer to sculpting as "weathering in fast forward."

Functionalism as an *ism* brings far too much to the table for it to claim the alleged virtues of metaphysical minimalism. The heart is a band of cardiac muscle and a bundle of neural structures (the bundle of His), divided up into chambers separated by valves. All this apparatus moves blood; it also makes sounds and gives rise to a palpable "pulse." It also allows for the widespread distribution of oxygen and the removal of carbon monoxide. It also results in thoracic sensations associated with emotion. It brings color to the cheeks and warmth to the extremities. If one takes the "function" of the heart to be coloration of the cheeks, then the heart and a pad of rouge in a cosmetics case serve the same function. Thus, if our wet brain can (as Dennett says) be replaced by silicon chips, just in case the "functions" served by the brain are now properly carried out, then the heart can be replaced by rouge, if the "function" of the heart is to color the cheeks—or by a clock, if the "function" of the heart is to make regular ticking sounds. If, indeed, it is a "function" of the brain to generate consciousness, there must be some mode of generation as well as an end state that is plausibly connected to what is taken to be the source. This is unproblematic when the end state is the movement of blood, the creation of regular sounds, the reddening of the cheeks. It is highly problematic when the end state is indubitably knowable solely to the possessor of that particular brain. This same epistemic authority would be retained by the possessor of the silicon chips just in case the chips generated the same end state. Even the designer of the robot would have no greater epistemic authority than Aunt Helen in declaring the creation to be conscious.

This is the well-known "first-person problem," and it has attracted predictable attention over the centuries and long before Descartes took his turn. I do not take it as a problem, however, for I take it to be basic. To make my own position clearer, let me quote some lines from David Chalmers, with whom I am in close agreement on some but not all aspects of these knotty issues. Chalmers has written:

> As I see it, the distinctive task of a science of consciousness is to systematically integrate two key classes of data into a scientific framework: *third-person data* about behavior and brain processes, and *first-person data* about subjective experience. When a conscious system is observed from the third-person point of view, a range of specific behavioral and neural phenomena present themselves. When a conscious system is observed from the first-person point of view, a range of specific subjective phenomena present themselves. I think both sorts of phenomena have the status of data for a science of consciousness.[3]

Is this either a distinctive task at all and, if so, is it the right task for "a science of consciousness"? One venerable specialty within medicine is clinical neurology, whose distinctive task is, indeed, connecting a patient's subjective reports to the known functions of the nervous system. Were it not for the facts gleaned by clinicians in neurology there would be little reason for focusing on the brain at all! We are taunted by something called the mind-brain problem owing to the observed alterations in mental function associated with brain pathologies. Not to be merely argumentative, I should say that dentists, too, have as one distinctive task connecting the first-person reports of pain (and associated pain behavior) to x-ray data on the state of the dental nerves. Any number of symptoms of diseases of the nervous system can be produced by vitamin B12 deficiency. Numbness in the hands may be sign of circulatory disorders, lactic acid formation from fatigue, muscle weakness, peripheral nerve damage, tumors in the brachial plexus, or anomalous conditions in the brain. The point, of course, is that many areas of medicine, dentistry, nursing, nutrition, physical therapy, and psychotherapy must attend to the relationship between "*third-person data* about behavior and brain processes, and *first-person data* about subjective experience."

I should think that relationships of this sort, important though they are and, as noted, venerably studied long before anyone ever thought of a "science of consciousness," have very little to do with such a science. The forms of life made possible by consciousness are only tangentially captured by terms such as "subjective experience," especially as these reach little beyond the rudiments of visual and auditory sensations. I should think that the distinctive task of a science of consciousness would be a credible and systematic account of the manner in which knowledge, desire, belief, and judgment come to be integrated into action plans by entities that have and that take an interest in themselves and in others. Put another way, the distinctive task pertains to what is distinctive about human life, which is not merely or primarily "subjective experience." What is distinctive about it is its amenability to rhetorical sources of motivation, to desires grounded in moral precepts, to forms of art and play, belief and conviction, and hopes and intuitions, by which "behavior" rises to the level of personal responsibility.

Bound up with these considerations is a life that is irreducibly *civic*: not simply "social" in the sense of patterns of mutual influence but *civic* in the sense of regulative precepts, rules of law, of etiquette, of ethics. Such a life requires *self*-consciousness as a precondition for imputing standing to others. Civic life obliges one to take the part of another, to see the world or the setting or the problem as another is seeing it. Such a life calls for *empathy*, which, in the end, is a form of co-consciousness, an achievement of a developed, refined, and disciplined consciousness. Only under these descriptions are whole classes of behavior (e.g., "criminal," "saintly," "heroic," "generous," "envious") intelligible as something other than muscle twitches.

The ancient world was not attentive to *our* problem of (or problem *with*) consciousness, for its two greatest philosophers were tentative in their ontologies, Plato rejecting materialism and Aristotle denying that natural science was "ultimate." They were not attentive to *our* problem of (or problem *with*) consciousness for they did not regard the essence of rationality and mentation to be "subjective experience." A psychology going little further in its inquiries than *phantasmata* would be taken as devoid of explanatory resources. They were not attentive to *our* problem of (or problem *with*) consciousness because they did not regard the central mission of philosophy to be the analysis of concepts. Analysis was understood as a tool, a method of clarifying words and notions so that substantive disagreements might be engaged. Within limits, their modes of analysis and their orientating assumptions were the gift of a folk psychology that has consciousness as its point of origin. It is that in virtue of which there might be problems of any sort!

Folk psychology in these respects just is the foundational science. I say this not to immunize village credulity and seasonless prejudice against the harsh light of scientific and logical assessment. To say that physics is the foundational science for all that is manifest in the form of matter and energy is not to say that physics at any point in its development is sound in its methods or correct in its core propositions. Identifying a discipline or field of inquiry as foundational is not an evaluation of its achievements but the identification of the scientific framework within which all facts of a given kind must find their systematic treatment. All the facts of consciousness, including those actions and experiences arising from it, have as the framework within which to be subjected to systematic treatment *folk psychology*. If that psychology is in a primitive state relative to its distinctive task, then it is that psychology that requires development and refinement, rather than being abandoned so that some other framework—for example, "cognitive neuroscience"—might be adopted with less frustration or embarrassment.

Does this conclusion force the tireless workers out of the laboratory and back into the armchair? The position one has as a vantage point must be chosen, not compelled. If Locke and Hume, Berkeley and Kant, Descartes and Plato (readers know how long a list might be composed here) regarded their own moral and intellectual lives as representative enough for the purpose of deriving a "science" of mental life, it would be rash to rule out their mode of inquiry. The prepared and serious mind reflecting on the nature of its own operations is not under some special burden of establishing its validity. Actually, it's the other way round: a third-person account of *my* toothache is what bears the burden, and it bears it credibly only when it matches up very well with most toothaches. In that case, if it fails in my case, there would be grounds of suspecting me or my nervous system as in some possibly interesting way eccentric.

How much progress might be expected of those in armchairs—how much progress might be expected of those in laboratories—depends chiefly on just who is in each location. If advice might be useful coming from one who has labored earnestly in both vineyards, that of philosophical reflection and brain science, I would suggest that the "cognitive revolution" be medicated a bit, calmed down, given a pause to regain its sobriety, so that its practitioners might recover a less lofty place but one worthy of praise for its energy and occasionally insightful findings; viz., functional neuroanatomy, chiefly within the context of clinical neurology. As for the philosophers, it would appear advisable to resist the temptation to hyphenate our endearing if vexing subject. Brain function and mental life are connected, to be sure. So, too, is kidney function and mental life. I would no more be inclined toward "neurophilosophy" than "hepatophilosophy." It is for others to end their declarative sentences with exclamation points. Ours should end with a semicolon;

NOTES

1. THE GREEKS (AGAIN) AND THE "CONSCIOUSNESS" PROBLEM

1. David Chalmers, for example, states that consciousness "poses the most baffling problems in the science of the mind. . . . There is nothing that is harder to explain." David Chalmers, "Facing up to the Problem of Consciousness," *Journal of Consciousness Studies* 2, no. 3 (1995): 200–219.

2. William James, *Principles of Psychology* (New York: Henry Holt, 1890).

3. Consider only the degree to which conscious and pressing aspirations influence what and how past events are recalled, and how "pseudo-reminiscence" and the personalized construction of the past express the demands on conscious life.

4. I will offer in this and the next chapter evidence against this view, a view defended in an interesting and influential essay by Wallace Matson, "Why Isn't the Mind-Body Problem Ancient?" in *Mind, Matter, and Method: Essays in Philosophy of Science in Honor of Herbert Feigl*, ed. Paul Feyerabend and Grover Maxwell (Minneapolis: University of Minnesota Press, 1996), 92–102.

5. Matson, "Why Isn't the Mind-Body Problem Ancient?"

6. Ibid., 100.

7. All citations are from *The Complete Works of Aristotle*, ed. Jonathan Barnes (Princeton, N.J.: Princeton University Press, 1984).

8. Aristotle, *On the Soul*, 408a10–15.

9. Ibid., 408b13–16.

10. Ibid., 408b30.

11. On this perennial question, an instructive and balanced essay is Philip van der Eijk, "Aristotle on the Soul-Body Relationship," in *Psyche and Soma: Physicians and Metaphysicians on the Mind-Body Problem from Antiquity to Enlightenment*, ed. John Wright and Paul Potter (Oxford: Clarendon Press, 2000), chap. 3.

12. Aristotle, *On the Soul*, 408b30.

13. Ibid., 412b5–10.

14. Ibid., 430a17–23.

15. Ibid., 428a.

16. Ibid., 429b18–25.

17. See especially 436b8–30.

18. Victor Caston, "Aristotle on Consciousness," *Mind* 111 (2002): 751–815.

19. Aristotle, *On the Soul*, 425b13–16.

20. Ibid., 3.2.436b22.

21. J. Noel Hubler, "The Perils of Self-Perception: Explanations of Apperception in the Greek Commentators on Aristotle," *Review of Metaphysics* 59, no. 2 (2005): 287–311.

22. Ibid., 290.

23. Aristotle, *On the Soul*, 3.3; 437a18–23. The "cognizance of what is" (γνωριζει τωνοντων) refers to what must be understood as *conscious* apprehension.

24. An especially economical expression of the "physics-is-complete doctrine" can be found in G. Helman and F. Thompson, "Physicalism: Ontology, Determination, Reduction," *Journal of Philosophy* 72 (1975): 551–564. They describe their project thus: "We seek to develop principles of *physical determination* that spell out rather precisely the underlying physicalist intuition that the physical facts determine all the facts."

25. Aristotle, *Metaphysics*, 9.1064b10–15. His "separate and immovable" is χωριστη και ακινητος, where χωριστη from χωρις is meant to imply "without the help or influence" of anything external to itself.

26. Ibid., 1072b10–15.

27. Ibid.

28. For additional examples, consider his discussion in *De Anima* beginning at 413a.

29. T.M. Robinson, "The Defining Features of Mind-Body Dualism in the Writings of Plato," in *Psyche and Soma: Physicians and Metaphysicians on the Mind-Body Problem from Antiquity to Enlightenment*, ed. John Wright and Paul Potter (Oxford: Clarendon Press, 2000), chap. 2.

30. Plato, *Laws*, 897E.

31. Ibid.

32. Ibid., 998A.

2. THE PROBLEM OF CONSCIOUSNESS "SOLVED"

1. Daniel Dennett, *Consciousness Explained* (London: Penguin, 1992).

2. Ibid., 260.

3. It was Catherine Wilkes who brought this to my attention, Trinity Term 2004.

4. Thomas Hobbes, *Leviathan* (1651), bk. 1, chap. 7; 31.

5. Gilbert Ryle, *The Concept of Mind* (Chicago: University of Chicago Press, 1949). Ryle's critique will be considered further in the next chapter.

6. The current status of theoretical physics leaves open the question of just what counts as "physical." Thus, the question remains open as to what a "completed" physics might come to include, not to mention the metaphysical question as to the criteria of "completeness." For example, must a "completed" physics include teleological conditionals?

7. See especially Carl Hempel, "Comments on Goodman's *Ways of Worldmaking*," *Synthese* 45 (1980): 193–199.

8. David Papineau, *Thinking About Consciousness* (Oxford: Oxford University Press, 2002), 253–254.

9. That the "completeness of physics" riposte has been overplayed in defenses of materialism is cogently argued by Tim Crane, "Mental Causation," *Proceedings of the Aristotelian Society*, suppl. vol. 69 (1996): 211–236.

10. Michael Lockwood, *Mind, Brain, and the Quantum* (Oxford: Blackwells, 1989).

11. William James, *Principles of Psychology* (New York: Henry Holt, 1890). See chapter 5, "The Automaton Theory."

12. See, for example, Frank Jackson, "Mental Causation," *Mind* 105 (1966): 377–413.

13. On this point, see D. N. Robinson, "Cerebral Plurality and the Unity of Self," *American Psychologist* 37 (1982): 904–910.

14. Consider "lucid dreaming," where one is aware that one is dreaming.

15. Gassendi's and Hobbes' criticisms and Descartes' replies are included in *The Philosophical Writings of Descartes*, ed. John Cottingham et al. (Cambridge: Cambridge University Press, 1998). Gassendi was an ordained priest and a leading advocate of Epicurean materialist psychology. He was famous in his day, a teacher of Moliere, a provost of the Cathedral at Digne, and holder of the Royal College chair in mathematics.

16. His *Treatise on Man* and other major works routinely explain perception, emotion, the formation of thoughts, etc. in physiological terms.

17. Morgan expressed his "Canon" thus: "In no case may we interpret an action as the outcome of the exercise of a higher psychical faculty, if it can be interpreted as the outcome of one which stands lower in the psychological scale." C. L. Morgan, *Introduction to Comparative Psychology*, 2nd ed., rev. (London: Walter Scott, 1903), 53.

18. An informative biographical sketch of Alexander is provided in John Slater's introduction to *Collected Works of Samuel Alexander* (Bristol: Thoemmes Press, 2000).

19. The first edition appeared in 1920. The passage here is taken from the second edition, (London: Macmillan, 1927), 5–6.

20. Jaegwon Kim, "Non-reductivism and Mental Causation," in *Mental Causation*, ed. John Heil and Alfred Mele (New York: Oxford University Press, 2000), 198.

21. It is difficult to conceive of an "emergent" property that would not be explicable in terms of the totality of properties of the constituents. Iron and oxygen, pure and simple, do not have the properties of ferrous oxide, but in which sense would a complete understanding of iron and of oxygen not include the properties arising from the right combination of the two?

22. Kim, "Non-reductivism and Mental Causation," 190–191.

23. Thomas Henry Huxley, "On the Hypothesis That Animals Are Automata and Its History," *Fortnightly Review* 22 (1874): 199–245.

24. Ibid.

25. Joseph Levine, "Conceivability, Identity, and the Explanatory Gap," in *Toward a Science of Consciousness III*, ed. Stuart R. Hameroff, Alfred W. Kaszniak, and David J. Chalmers (Cambridge, Mass.: MIT Press, 1999). The early version of Levine's position is Joseph Levine, "Materialism and

Qualia: The Explanatory Gap," *Pacific Philosophical Quarterly* 64 (1983): 354–361.

26. David Chalmers, *The Conscious Mind: In Search of a Fundamental Theory* (New York: Oxford University Press, 1996).

27. David Chalmers, "Facing up to the Problem of Consciousness," *Journal of Consciousness Studies* 2, no. 3 (1995): 200–219.

28. Dennett, *Consciousness Explained.*

29. Of course, considering something in "functional" terms does not justify begging the question of materialism.

30. For a classical statement of mind-brain identity, see D. M. Armstrong, *A Materialist Theory of the Mind* (London: Routledge, 1968). A more recent defense of type identity and a critical appraisal of alternatives is U. T. Place, "Token- Versus Type-identity Physicalism," *Anthropology and Philosophy* 2, no. 2 (1999).

31. Surely the most influential early statement of the functionalist position was Hilary Putnam's "Psychological Predicates," in *Art, Mind, and Religion*, ed. W. H. Capitan and D. D. Merrill (Pittsburgh, Penn.: University of Pittsburgh Press, 1967). The more frequently cited source is Putnam's "The Nature of Mental States," in his *Mind, Language, and Reality* (Cambridge: Cambridge University Press, 1975).

32. Daniel Dennett, *Brainstorms* (Cambridge, Mass.: MIT Press, 1978).

33. Dennett, *Consciousness Explained.*

34. For a recent version of Block's position, see Ned Block, "Semantics, Conceptual Role," in *The Routledge Encyclopedia of Philosophy* (London: Routledge, 1997). Supporting and converging arguments for CRS can be found in G. Miller and P. Johnson-Laird, *Language and Perception* (Cambridge, Mass.: MIT Press, 1976); and C. Peacocke, *A Theory of Concepts* (Cambridge, Mass.: MIT Press, 1992) (A CRS-oriented account of the nature of concepts).

35. The first statement of the argument is John Searle, "Minds, Brains, and Programs," *Behavioral and Brain Sciences* 3 (1980): 417–424; and "Intrinsic Intentionality," *Behavioral and Brain Sciences* 3 (1980): 450–456. See also his *Minds, Brains, and Science* (Cambridge, Mass.: Harvard University Press, 1984).

36. Ned Block, "Semantics, Conceptual Role."

37. Thomas Nagel, "What Is It Like to Be a Bat?" *Philosophical Review* 83 (1974): 435–450. I will consider this again in the next chapter.

38. Ibid, 435.

39. Ibid.

40. David Armstrong, "What Is Consciousness?" in *The Nature of Mind* (Ithaca, N.Y.: Cornell University Press, 1981).

41. One might point to the reliable relationship between increases in perceived brightness and increases in the firing rate of fibers in the optic nerve, thus suggesting that a "type" of phenomenological change perfectly matches a "type" of physical state or process. But the general rule finds the objects of perception strikingly stable (e.g., size, color, and shape constancy) under conditions in which the physical aspects of the stimulus undergo significant change.

3. "CARTESIANISM" REVISITED

1. Peter King, *Forming the Mind* (Berlin: Springer Verlag, 2005), 1.

2. Gerard O'Daly, *Augustine's Philosophy of Mind* (Los Angeles: University of California Press, 1987), 80.

3. Excellent treatments of the authoritative works produced from antiquity to the Enlightenment are found in John Wright and Paul Potter, eds., *Psyche and Soma: Physicians and Metaphysicians on the Mind-Body Problem from Antiquity to Enlightenment* (Oxford: Clarendon Press, 2000).

4. From his *Preface to Divine Dialogues* and quoted in R. L. Brett, *The Third Earl of Shaftesbury* (London: Hutchinson's University Library, 1951), 25. Brett's study of Shaftesbury's philosophy remains revealing and authoritative. It is a pity it is not more widely cited. I have benefited from his insights into the quandary Descartes created for the aesthetic and religious commentators of the period.

5. Jaegwon Kim, "The Non-Reductivist's Troubles with Mental Causation," in *Mental Causation*, ed. John Heil and Alfred Mele (Oxford: Clarendon Press, 2000), 189. Italics in the original.

6. Ibid., 190, italics in the original.

7. Hobbes, *Leviathan* (1651), bk. 1, chap. 1.

8. Samuel Mintz, *The Hunting of Leviathan: Seventeenth-Century Reactions to the Materialism and Moral Philosophy of Thomas Hobbes* (Cambridge: Cambridge University Press, 1970), vi.

9. Hobbes's physics required a *plenum* such that, in principle, there could be no vacuum. He insisted on this even as Boyle succeeded with his

air pump to create partial vacuums in the laboratory. A close account of their animated exchanges is Steven Shapin and Stephen Schaffer, *Leviathan and the Air-Pump: Hobbes, Boyle, and the Experimental Life* (Princeton, N.J.: Princeton University Press, 1985). See Mintz, *The Hunting of Leviathan*, 87, for a discussion of Boyle's published attack (*An Examen of Mr T. Hobbes* [1662]) on Hobbes as scientist.

10. Pierre Simon Laplace, introduction to *A Philosophical Essay on Probabilities* (New York: Wiley and Sons, 1902). Translated from the 6th French ed. (1812).

11. Shaftesbury, *Characteristics of Men, Manners, Opinions, Times*, ed. Lawrence E. Klein (Cambridge: Cambridge University Press, 1999), 304.

12. Shaftesbury, *Preface to Divine Dialogues*, quoted in Brett, *The Third Earl of Shaftesbury*, 27.

13. Heinrich von Staden, "Body, Soul, and Nerves: Epicurus, Herophilus, Erisastratus, the Stoics, and Galen," in *Psyche and Soma*, ed. John P. Wright and Paul Potter, 79.

14. St. Augustine, *De Trinitate*, 15.21.

15. "Academic" here refers to latter-day philosophers, skeptical and materialistic in ways that defied the positions developed in Plato's Academy.

16. Descartes' reliance on or at least compatibility with Augustine's "*Cogito*" was noted as far back as Pascal, who credits Augustine with priority but Descartes with recognizing the fuller implications of the argument. See Blaise Pascal, "De l'Esprit geometrique," in *Les Provinciales, Pensees et Opuscules*, ed. Gerard Ferryrolles and Phillipe Sellier (Paris, 2004), 142.

17. St. Augustine, *De Trinitate*, 15.21.

18. Ibid., chap. 11.

19. Anthony Kenny, *Aquinas on Mind* (New York: Routledge, 1994), 16–17.

20. Descartes, *Discourse on Method*, in *The Philosophical Writings of Descartes*, trans. John Cottingham, Robert Stoothoff, and Dugald Murdoch (Cambridge: Cambridge University Press, 1985), 1:139–140. All quotations from Descartes are taken from this source.

21. On this point there is room for interpretive disagreements, but the favored position is one that spares Descartes the title of class dunce!

22. For an instructive account of this curriculum and its purposes, see Allen P. Farrell, S.J., *The Jesuit Code of Liberal Education: Development and Scope of the* Ratio Studiorum (Milwaukee: Bruce Publishing Co., 1938).

23. Desmond Clarke, *Descartes' Theory of Mind* (Oxford: Clarendon Press, 2003).

24. Ibid., 14.

25. Descartes, *Discourse on Method*, 1:114–115.

26. Ibid., 1:114.

27. Ibid.

28. Descartes, *Principles of Philosophy*, 1:184.

29. Charles Siewert, *The Significance of Consciousness* (Princeton, N.J.: Princeton University Press, 1998), 51.

30. For a detailed and illuminating study of the manner in which Descartes understood Aristotle's theory and his less than adequate attempt to improve or refute it, see Sarah Byers, "Life as 'Self-Motion': Descartes and 'The Aristotelian' on the Soul as the Life of the Body," *Review of Metaphysics* 59 (2006): 723–755.

31. Descartes, *Discourse on Method*, 1:46–50.

32. Byers, "Life as 'Self-Motion,'" 723.

33. A most informing treatment of the entire matter is Paul Livingston, *Philosophical History and the Problem of Consciousness* (Cambridge: Cambridge University Press, 2004).

34. Ludwig Wittgenstein, *Philosophical Investigations* (Oxford: Blackwells, 1974), 1:293.

35. By "new" here, I would be understood as marking the decisive moment with Hume's influential philosophy. It goes without saying that developments such as those under consideration have no fixed dates of birth.

36. Livingston, *Philosophical History and the Problem of Consciousness*, 110.

37. Reid's discussion is found in chapter 6, section 24. Thomas Reid, *An Inquiry Into the Human Mind on the Principles of Common Sense* (1764), ed. Derek Brookes (State Park: Penn State University Press, 1997). For a discussion of this, see Rom Harré and Daniel N. Robinson, "What Makes Language Possible? Ethological Foundationalism in Reid and Wittgenstein," *Review of Metaphysics* 50 (1997): 483–498.

38. Richard Rorty, *Philosophy and the Mirror of Nature* (Princeton, N.J.: Princeton University Press, 1979).

39. Ibid., 5.

40. I refer to Thomas Reid's *An Inquiry Into the Human Mind* (1764)

and, in relation to Wittgenstein, especially to Reid's discussion of "natural language" and his broadly pragmatic criteria over and against speculative modes of understanding. Rorty discusses aspects of Reid's philosophy a half dozen times, but not on this point.

41. Rorty, *Philosophy and the Mirror of Nature*, 5.

42. Wilhelm Windelband's "Geschichte und Naturwissenschaft" was published in 1915. Here Windelband distinguishes between historical and natural-science modes of explanation. Along with Brentano, Wundt, Dilthey, and many others, Windelband reinforced the very outlook and arguments for which Heidegger is given far too much credit and on which his own education in philosophy was based. Both Rousseau and Herder advanced theories on the nature and origin of language that anticipated the developments Rorty has in mind. In this connection, see John H. Moran and Alexander Gode, trans., *On the Origin of Language: Two Essays. Jean-Jacques Rousseau, Johann Gottfried Herder* (Chicago: University of Chicago Press, 1986).

43. The authoritative study is Johannes Fritsche, *Historical Destiny and National Socialism in Heidegger's Being and Time* (Berkeley: University of California Press, 1999). Fritsche makes clear that *Being and Time* is entirely consistent with prevailing right-wing treatises from prewar Germany. Of course, one can concur with a theory of culturally driven "inevitabilities" and the authenticity of the morality of the *Volk* without accepting the political programs conceptually consonant with that theory.

44. This is argued in Rorty, *Philosophy and the Mirror of Nature*, 22–23.

45. Rorty, *Philosophy and the Mirror of Nature*, 26.

46. T. Crane, "Intentional Objects," *Ratio* 14, no. 4 (December 2001): 336–349.

47. Gilbert Ryle, *The Concept of Mind* (Chicago: University of Chicago Press, 1949). Here is the *locus classicus* of the war on "Cartesianism," though Ryle at least is aware of the long and distinguished pedigree of the two-substance ontology and its reliance on versions of the *Cogito*.

48. John Dewey, "Soul and Body," *Bibliotheca Sacra* 43 (1886): 242.

49. Descartes, *The Passions of the Soul*, 1:339.

50. Readers will be reminded here of Andy Clark and David Chalmers, "The Extended Mind," *Analysis* 58 (1998): 7–19, where an "extended mind" hypothesis includes not just brain but also body and the external physical world as entering into the ontology of the mental.

51. An instructive discussion of the entire affair is developed by Andrew Feffer, "The Presence of Democracy: Deweyan Exceptionalism and Communist Teachers in the 1930s," *Journal of the History of Ideas* 66 (2005): 79–97.

52. Robert Audi, "Mental Causation: Sustaining and Dynamic," in Heil and Mele, eds., *Mental Causation*, 53.

53. The best source is Donald Davidson, *Essays on Actions and Events* (Oxford: Oxford University Press, 1980). Davidson has revisited his theory often, with measured gains in clarity.

54. Davidson's clearest summary is in his "Mental Events," in *Philosophy of Mind*, ed. John Heil (Oxford: Oxford University Press, 2004), chap. 39.

55. Ibid., 686.

56. Ibid., 694.

57. A more recent account of this is Donald Davidson, "Thinking Causes," in Heil and Mele, eds., *Mental Causation*, 3–18.

58. A recent and informing resurrection of Descartes' arguments and his replies to such critics as Arnauld is Joseph Almog, *What Am I? Descartes and the Mind-Body Problem* (New York: Oxford University Press, 2002).

4. HIGHER-ORDER THOUGHT: A MACHINE IN THE GHOST

1. Representative here is David Rosenthal, "A Theory of Consciousness," in *The Nature of Consciousness: Philosophical Debates*, ed. Ned Block, Owen Flanagan, and Güven Güzeldere (Cambridge, Mass.: MIT Press, 1997), 729–753. Rosenthal writes within the larger community of reductionist theorists who would have all such "states" understood finally as brain states, but nothing in the argument of HOT requires or establishes any such thing.

2. Alvin Goldman, "Consciousness, Folk Psychology, and Cognitive Science," *Consciousness and Cognition* 2 (1993): 366.

3. Descartes, *Discourse on Method*, in *The Philosophical Writings of Descartes*, trans. John Cottingham, Robert Stoothoff, and Dugald Murdoch (Cambridge: Cambridge University Press, 1985), 1:56. All quotations from Descartes are taken from this source.

4. Charles Siewert, *The Significance of Consciousness* (Princeton, N.J.:

Princeton University Press, 1998), 196. Chapter 6 of this book, "Consciousness and Self-reflection," is an especially well-developed critique of the entire HOT approach and orienting assumptions.

5. For a developed version of HOT-d, see Peter Carruthers, *Phenomenal Consciousness: A Naturalistic Theory* (Cambridge: Cambridge University Press, 2000). For the HOT-nd variety, see Daniel Dennett, "Toward a Cognitive Theory of Consciousness," in *Perception and Cognition: Issues in the Foundations of Psychology*, ed. C. Savage, Minnesota Studies in the Philosophy of Science 9 (1978).

6. Carruthers, *Phenomenal Consciousness*, chap. 5.

7. For a critique of the usual forms of representationalism, see Charles Travis, "The Silence of the Senses," *Mind* 113 (2004): 57–94.

8. The originating source is L. Weiskrantz, *Blindsight* (Oxford: Oxford University Press, 1986). The phenomenon itself would have caused no surprise to specialists in psychophysics and Signal Detection Theory (SDT). If the methods of SDT had been employed, it is likely that the "blind" patients would have been classified as seriously visually impaired. The really blind, under SDT conditions, would have detection rates at chance levels under all payoff conditions and at all levels of encouraged guessing. For an introduction, see T. D. Wickens, *Elementary Signal Detection Theory* (New York: Oxford University Press, 2002).

9. William James, *Principles of Psychology* (New York: Henry Holt, 1890), 1:165–166.

10. Tom Nagel, "What Is It Like to Be a Bat?" *Philosophical Review* 83 (1974).

11. David Chalmers, "Consciousness and Its Place in Nature," in *Philosophy of Mind: Classical and Contemporary Readings*, ed. David Chalmers (Oxford: Oxford University Press, 2002).

12. Luis Bermudez, *Thinking Without Words* (Oxford: Oxford University Press, 2003).

13. Herbart's *Lehrbuch* went through several editions and was translated into English. Fechner offers this appraisal in his introductory remarks in his *Element der Psychophysik* (1860), translated by Helmut Adler as *Elements of Psychophysics.*

14. Fred Dretske, *Naturalizing the Mind* (Cambridge, Mass.: MIT Press, 1995).

15. Attempts to ground consciousness in objective, nonverbal criteria

have been developed within the brain sciences over a course of years. For a recent review and for currently favored criteria, see Anil Seth, Bernard Baars, and David Edelman, "Criteria for Consciousness in Humans and Other Animals," *Consciousness and Cognition* 14 (2005): 119–139. The electrical patterns of low-amplitude, fast discharges within thalamocortical paths, characteristic of human consciousness, are well developed in mammals.

16. G. E. Moore, *Principia Ethica* (Cambridge: Cambridge University Press, 1903), 7ff.

17. David O. Brink, "Realism, Naturalism, and Moral Semantics," in *Moral Knowledge*, ed. E. F. Paul, F. D. Miller, and J. Paul (New York: Cambridge University Press, 2001), 155.

18. William James, "Does 'Consciousness' Exist?" *Journal of Philosophy, Psychology, and Scientific Methods* 1 (1904): 477–491.

19. Ibid., 477.

20. Ibid., 478.

21. Ibid., 479.

22. Daniel N. Robinson, *Praise and Blame: Moral Realism and Its Applications* (Princeton, N.J.: Princeton University Press, 2002).

5. SELF-CONSCIOUSNESS

1. Thomas Metzinger, *Being No One: The Self-Model Theory of Subjectivity* (Cambridge, Mass.: MIT Press, 2003), 1.

2. G. E. M. Anscombe, "The First Person," in *Mind and Language*, ed. Samuel Guttenplan (Oxford: Clarendon Press, 1975), 45–65.

3. Ibid., 62.

4. Ibid., 51.

5. Hilary Putnam, "The Meaning of 'Meaning,'" in *Language, Mind, and Knowledge*, ed. K. Gunderson (Minneapolis: University of Minnesota Press, 1975).

6. Tyler Burge, "Individualism and the Mental," in *Midwest Studies in Philosophy* IV, ed. Peter French, Theodore Uehling, and Howard Wettstein (Minneapolis: University of Minnesota Press, 1979), 73–121.

7. Descartes, *Discourse on Method*, in *The Philosophical Writings of Descartes*, trans. John Cottingham, Robert Stoothoff, and Dugald Murdoch (Cambridge: Cambridge University Press, 1985), part 2. All quotations from Descartes are taken from this source.

8. The most precise study is C.H. Graham and Yun Hsia, "Studies of Color Blindness: A Unilaterally Dichromatic Subject," *Proceedings of the National Academy of Sciences* 45 (1959): 96–99.

9. Frank Jackson, "What Mary Didn't Know," *Journal of Philosophy* 83 (1986): 291–295.

10. Ibid.

11. Bertrand Russell's well-known distinction is developed in chapter 5 of *The Problems of Philosophy* (Oxford: Oxford University Press, 1912).

12. Several of these pages are drawn from a lecture given at Aarhuus University, Denmark, in the fall of 2004. I thank the attending faculty and students for their interesting and helpful questions and comments.

13. The seminal modern contributions to the issue are David Wiggins, *Identity and Spatio-Temporal Continuity* (Oxford: Blackwell, 1967); and his *Sameness and Substance* (Oxford: Blackwell, 1980). Especially influential has been Derek Parfit, *Reasons and Persons* (Oxford: Oxford University Press, 1984).

14. Thomas Hobbes, *De Corpore*, in *The English Works of Thomas Hobbes*, ed. Molesworth (1839), 1:136.

15. Ibid.

16. John Locke, *An Essay Concerning Human Understanding*, ed. Peter Nidditch (Oxford: Clarendon Press, 1975).

17. Ibid.

18. Ibid., 460.

19. Robert Boyle, "Some Physico-Theological Considerations About the Possibility of the Resurrection," in *Selected Philosophical Papers of Robert Boyle*, ed. M. A. Stewart (New York: Manchester University Press, 1979), 198.

20. The best source for the exchange of letters and for a full discussion of the issues arising from Locke's theory is G. A. J. Rogers, ed., *The Philosophy of Edward Stillingfleet; Including His Replies to John Locke*, 6 vols. (Bristol: Thoemmes Publishing Co., 1999). Volume 5 contains *The Bishop of Worcester's Answer to Mr Locke's Letter, Concerning some Passages Relating to his Essay of Humane Understanding* (1697, 158 pages), bound with *The Bishop of Worcester's Answer to Mr Locke's Second Letter wherein his Notion of Ideas is Prov'd to be Inconsistent with Itself* (1698, 186 pages). .

21. Locke, *An Essay Concerning Human Understanding*, 385.

22. Alexander Bird, *Philosophy of Science* (Montreal: McGill-Queen's University Press, 1998), 96–97.

23. Saul Kripke, *Naming and Necessity* (Cambridge, Mass.: Harvard University Press, 1980), 32–36.

24. For Nancy Cartwright's version, see *How the Laws of Physics Lie* (Oxford: Oxford University Press, 2002) and *The Dappled Universe* (Cambridge: Cambridge University Press, 1999). See also Bastian van Fraassen, *The Scientific Image* (Oxford: Clarendon, 1980).

25. H. Clark Barrett, "On the Functional Origins of Essentialism," *Mind and Society* 2 (2001): 1–30.

26. S. A. Gelman and J. D. Coley, "The Importance of Knowing a Dodo Is a Bird," *Developmental Psychology* 26 (1990): 796–804.

27. Barrett, "On the Functional Origins of Essentialism," 5.

28. David Hume, *A Treatise of Human Nature*, book 1, part 4, section 6.

29. Ibid.

30. Ibid.

31. Derek Parfit, *Reasons and Persons*.

32. David Wiggins, *Sameness and Substance Renewed* (Cambridge: Cambridge University Press, 2001).

33. Parfit, *Reasons and Persons*, 205.

34. Ibid., 220.

35. Ibid., 211.

36. Ibid., 325.

37. Ibid., 327. I hesitate to think of the bearing all this might have on the law of contracts; nor is it obvious that anyone else would in the circumstance be bound by agreements entered into with our "new" old man when once he was young. Though lacking in rigorous analyticity, demonstrations of the nearly comical implications of a philosophical thesis might at least encourage further adumbrations of its merits.

38. Christine Korsgaard, "Personal Identity and the Unity of Agency: A Kantian Response to Parfit," *Philosophy and Public Affairs* 18 (1989).

39. See Christopher Fox, *Locke and the Scriblerians* (Berkeley: University of California Press, 1988).

40. Thomas Reid, *Essays on the Intellectual Powers of Man* (Edinburgh: 1785), 319–320. From essay 3, chap. 4, "Of Identity."

41. Ibid.

42. John Hawthorne, "Why Humeans Are out of Their Minds" *Nous* 38, no. 2 (2004): 351–358.

43. Ibid., 351.

44. Ibid.

45. Ibid.

46. Illustratively, see Paul Boghossian, "The Transparency of Mental Content," in *Philosophical Perspectives*, ed. J. Tomberlin (Atascadero, Calif.: Ridgeview Press, 1994), 33–50.

47. No position is taken here on the more general internalist-externalist debate, which has to do with the warrants sufficient to qualify a state as one of knowing. Internalists would require of any knowledge-belief that the believer have cognitive access to the necessary ingredients of the knowledge. For externalists, some of the ingredients may not fall within the ambit of the believer's cognition.

48. The following material is adapted from my "Dissociation and the Foundations of Cognitive Psychology in Nineteenth-Century France," presented at the 102nd Annual Convention of the American Psychological Association (August 1994).

49. Alfred Binet, *On Double Consciousness* (1890), in *Significant Contributions to the History of Psychology*, ed. D. N. Robinson (Westport, Conn.: Greenwood Publishing, 1976–1977), series C, vol. 5.

50. Pierre Flourens, *Phrenology Examined* trans. C. Meigs (Philadelphia: Hoagland and Thompson, 1846), 123.

51. T. Ribot, *Diseases of Personality* (1891), in *Significant Contributions to the History of Psychology*, series C, vol. 1, p. 55.

52. Ibid., 55.

53. Ibid., 54.

54. Ibid., 56–57.

55. Ibid., 127.

56. Alfred Binet, *Alterations of Personality* (1896), in *Significant Contributions to the History of Psychology*, series C, vol. 1, p. 347.

57. Ibid., 281.

58. Oliver Sacks, *The Man Who Mistook His Wife for a Hat* (New York: Harper and Row, 1970), revived the excitement created earlier in the neurology clinics of Paris.

59. William James, *Principles of Psychology* (Cambridge, Mass.: Harvard University Press, 1981), 314. Originally published in 1890.

60. William James, "Does Consciousness Exist?" in *Essays in Radical Empiricism* (New York: Longman Green and Co., 1912), 1–4.

61. William James, *Principles of Psychology*, 371.

62. Jonardon Ganeri, "An Irrealist Theory of Self," *Harvard Review of Philosophy* 12 (2004): 61–80.

63. Ibid., 61.

64. See, for example, Kenneth Inada and Nolan Jacobson, eds., *Buddhism and American Thinkers* (Albany: State University of New York Press, 1984).

65. William James, "Does Consciousness Exist?", 38.

66. Ludwig Wittgenstein, *Philosophical Investigations* (Oxford: Blackwells, 1974), 1:404.

67. Marvin Minsky, "Minds Are Simply What Brains Do," *Truth: A Journal of Modern Thought* 2 (1988): 11–18.

68. R. Harré, "Language Games and Texts of Identity," in *Texts of Identity, ed. J. Shotter and K. Gergen* (London: Sage Publications, 1988), 22.

69. Ibid., 23.

70. Richard Rorty, "Solidarity or Objectivity?", in *Post-Analytic Philosophy*, ed. J. Rajchman and C. West (New York: Columbia University Press, 1985), 12.

71. William James, *Principles of Psychology*, 365.

72. P. Heelas and A. Locke, *Indigenous Psychologies: The Anthropology of the Self* (London: Academic Press, 1981).

73. William James, *Principles of Psychology*, 321.

74. Ibid.

75. Ibid., 324.

76. Immanuel Kant, *Critique of Pure Reason* [1787], trans. N. K. Smith (New York: St. Martin's Press, 1965), 333–334.

77. Ibid., 334.

78. Ibid., 28.

79. Ibid., 28–29.

6. EMOTION

1. M. R. Bennett and P. M. S. Hacker, *Philosophical Foundations of Neuroscience* (Oxford: Blackwell, 2003).

2. Daniel Bernoulli, "Exposition of a New Theory on the Measurement of Risk," *Econometrica* (1954): 22, 23–36. Originally published in 1738.

3. Illustrative is the work of Daniel Kahnemann, culminating in the Nobel Prize. See, e.g., D. Kahnemann and A. Tversky, "Choices, Values, and Frames," *American Psychologist* 39 (1984): 341–350.

4. Paul Ekman, *Emotions Revealed* (New York: Henry Holt, 2003).

5. Charles Darwin, *The Expression of the Emotions in Man and Animals* (London: John Murray, 1872).

6. Jaak Panksepp, *The Foundations of Human and Animal Emotions* (New York: Oxford University Press, 1998).

7. It is to Panksepp's credit that, having argued for human and animal emotions falling along a continuum—and thus giving credibility to the view that many species are surely endowed with feelings—he acknowledges the ethical burdens borne by those who expose such animals to conditions of captivity, distress, pain, and isolation. At the end of the day, however, the mission of science seems to overtake such scruples.

8. See John Cooper, *Reason and Emotion* (Princeton, N.J.: Princeton University Press, 1999), 464ff; and Richard Sorabji, *Emotion and Peace of Mind: From Stoic Agitation to Christian Temptation* (Oxford: Oxford University Press, 2000), chaps. 1 and 11.

9. Antony Kenny, *Action, Emotion, and Will* (London: Routledge, 1963); Robert C. Solomon, *The Passions: The Myth and Nature of Human Emotion* (Garden City, N.Y.: Anchor/Doubleday, 1976). In this same connection, see William Lyons, *Emotion* (Oxford: Oxford University Press, 1980).

10. Peter Goldie, *The Emotions* (Oxford: Oxford University Press, 2000).

11. See William James, "What Is an Emotion?" *Mind* 9: 188–205, for the sharpest development of his position.

12. The full account is in Antonio Damasio, *The Feeling of What Happens: Body and Emotion in the Making of Consciousness* (New York: Harcourt Brace, 1999).

13. Ibid., 280ff.

14. Ibid., 282.

15. Antonio Damasio, *Descartes' Error: Emotions, Reason, and the Human Brain* (New York: Avon Books, 1994).

16. All references to *The Passions of the Soul* are to the translation

by Stephen H. Voss (Indianapolis, Ind.: Hackett Publishing Company, 1989).

17. Ibid., Article 52.

18. Ibid., Article 212.

19. Donald Norman, *Emotional Design: Why We Love (or Hate) Everyday Things* (New York: Basic Books, 2004).

20. See in this connection Anthony Kenny, *Action, Emotion, and Will* (London: Routledge and Kegan Paul, 1963); Robert Solomon, "Emotions and Choice," in *Explaining Emotions*, ed. Amélie Rorty (Los Angeles: University of California Press, 1980), 251–281; Justin Oakley, *Morality and the Emotions* (London: Routledge and Kegan Paul, 1992).

21. Thomas Gilovich, *How We Know What Isn't So: The Fallibility of Human Reason in Everyday Life* (New York: Free Press, 1991).

7. MOTIVES, DESIRES, AND FULFILLMENT

1. G. E. M. Anscombe, "Modern Moral Philosophy," *Philosophy* 33 (1958): 1–19.

2. See Roger Crisp and Michael Slote, who provide an informing discussion of this in their introduction to *Virtue Ethics* (Oxford: Oxford University Press, 1997).

3. Aristotle, *Rhetoric*, bk. 1, chap. 1, 1355a. All citations and references are from *The Complete Works of Aristotle*, ed. Jonathan Barnes, trans. W. Rhys Roberts (Princeton, N.J.: Princeton University Press, 1984), vol. 2.

4. Ibid., bk. 1, chap. 2, 1358a.

5. Eugene Garver, *Aristotle's Rhetoric: An Art of Character* (Chicago: University of Chicago Press, 1994).

6. Garver develops this in his chapter 5. It is in chapter 5 of Hume's *Enquiry Concerning the Principles of Morals* that the link between utility and pleasure is considered.

7. Garver, *Aristotle's Rhetoric*, 151–156.

8. Aristotle, *Rhetoric*, bk. 1, chap. 1, 1358a 9–11.

9. Ibid., 1354a 20–27.

10. Ibid., bk. 1, chap. 2, 1357a 2–3.

11. Ibid., bk. I, chap. 5, 1360b 4–17.

12. Ibid., bk. I, chap. 8, 1366a 13–15.

13. Ibid., bk. I, chap. 7, 1365a 37.

14. Plato, *Menexenus*, 235D.

15. Aristotle, *Rhetoric*, bk. 1, chap. 14, 1374b 24–25.

16. Ibid., bk. 1, chap. 10, 1369a 5–6.

17. Ibid., bk. 1, chap. 13, 1374b 15–17.

18. Ibid., bk. 2, chap. 9, 1386b.

19. Ibid., bk. 2, chap. 1, 1378a 32.

20. Ibid., bk. 2, chap. 11, 1388b 32ff.

21. Ibid., bk. 2, chap. 12, 1389b 25–26.

22. Ibid., bk. 2, chap. 14, 1390b.

23. Ibid., bk. 2, chap. 15,1390b 21ff.

24. Descartes, *Meditations on the First Philosophy—Objections and Replies*. In *The Philosophical Writings of Descartes*, ed. John Cottingham et al. (Cambridge: Cambridge University Press, 1996), 2:95–96.

25. Anthony Kenny, ed. and trans., *Descartes' Philosophical Letters* (Oxford: Clarendon Press, 1970), 138.

26. Ibid.

27. Ibid.

28. Ibid.

29. Ibid.

30. Ibid., 140–144.

31. Ibid., 141.

32. John Heil, "Multiple Realizability," *American Philosophical Quarterly* 36 (1999): 189–208.

33. Kenny, *Descartes' Philosophical Letters*, 106, Letter to Marsenne, July 1641.

34. Ibid.

35. Ibid., 143.

36. Jim Crane, "The Mental Causation Debate," *Proceedings of the Aristotelian Society Supplementary Volume* 69 (1995).

37. Ibid., 1.

38. Tyler Burge, "Mind-Body Causation and Explanation," in *Mental Causation*, ed. John Heil and Alfred Mele (Oxford: Oxford University Press, 2000), 98.

39. Ibid., 104.

40. Representative of his position is David Lewis, "An Argument for the Identity Theory," in his *Philosophical Papers* (Oxford: Oxford University Press, 1983), vol. 1.

41. Ibid.

42. Letter to Princess Elizabeth, in Kenny, *Descartes' Philosophical Letters*, 139.

43. David Hume, *An Enquiry into the Human Understanding* (1748), sec. 4, pt. 1.

44. Ibid.

45. Ibid., sec. 7, pt. 1.

46. Ibid.

47. Ibid., sec. 5, pt. 1.

48. Mackie develops this in *The Cement of the Universe* (Oxford: Clarendon Press, 1974).

49. Ibid., 178.

50. A close analysis of Reid's critique of Hume and alternative position is Gideon Yaffe, *Manifest Activity: Thomas Reid's Theory of Action* (Oxford: Oxford University Press, 2004).

51. When Hume admits of the inability to give the "ultimate reason" for a universal "propensity," he can be as readily claimed as Reid by the school of intuitionists.

52. Thomas Reid, *An Inquiry into the Human Mind on the Principles of Common Sense* (1764), chap. 6, p. 196. All references are to the Pennsylvania State University Press edition (1997) edited by Derek Brookes.

53. Thomas Reid, *Essays on the Active Powers of the Human Mind* (1785), essay 1, chap. 2.

54. Thomas Reid, *Essays on the Intellectual Powers of Man*, ed. Derek Brookes (Edinburgh: Edinburgh University Press, 2002), essay 6, chap. 6, p. 499. Originally published in 1785.

55. Ibid., essay 2, chap. 4, p. 87.

56. William James, "The Dilemma of Determinism," in *The Will to Believe and Other Essays in Popular Philosophy* (New York: Henry Holt & Co., 1897), 148.

8. PLANS: AN EPILOGUE

1. Daniel Dennett, "The Zombic Hunch: Extinction of an Intuition?" Millennial Lecture, Royal Institute of Philosophy, November 28, 1999.

2. See, for example, Peter Marton, "Zombies Versus Materialists:

The Battle for Conceivability," *Southwest Philosophy Review* 14 (1998): 131–138.

3. David Chalmers, "How Can We Construct a Science of Consciousness?" in *The Cognitive Neurosciences III*, ed. M. Gazzaniga (Cambridge, Mass.: MIT Press, 2004).

INDEX

abstraction, 59–60
academic philosophers, 57–58
affect, decision making and, 147–48
Alas—Poor Darwin (Rose and Rose), 161
Alexander, Samuel, 29
algorithmic problem, 11–13
Anaxagoras, 8
anger, 179–80
animals: bat, experience of, 44–45, 91; consciousness present in, 59; epiphenomenalist view, 33–34; HOT theories and, 85, 89, 92–93, 95; self-movement of, 67–68; species-specific conditions, 92–93; subjectivity of, 44–45
anomalous monism, 78–81
Anscombe, Elizabeth, 102–3, 169–71
antiessentialism, 113
anti-individualism, 104–11
antimaterialism, 34–35
appearances, 143–44
Aquinas, Thomas, 59
Aristotle, 2, 3–5; emotions, treatment of, 158–59; essence, view of, 116–17; intellect as immovable, 4–7, 212n25; mind-body problem and, 2–12; moral philosophy and, 169–71; perception, view of, 7–8, 10–11; rejection of harmony theory, 4–6; *Works: Rhetoric*, 159, 171–82; *On the Soul*, 4
Armstrong, David, 44–45
association, laws of, 122–24
attention, 202
Audi, Robert, 78
audience, character of speaker and, 172–77
Augustine, 57–58, 217n16
authentic action, 25
awareness, 10, 83, 86, 89, 94, 98. *See also* self

Bain, Alexander, 36
bare consciousness, 71

Barrett, H. Clark, 120
bat, experience of, 44–45, 91
beetle in the box example, 69–70
beliefs, 8, 95, 131; identity and, 49–50
Bennett, Max, 146
Bernoulli, Daniel, 147
bifurcating conception of reality, 32
Binet, Alfred, 131–32, 133–34
Bird, Alexander, 117
blindsight, 89
Block, Ned, 42–43
Boyle, Robert, 54, 114–15
brain function: computer model, 39–40, 41, 42–43; consciousness as, 27–28; dependent on mental states, 23–24; emotion attributed to, 150–51; higher functions of, 88; modular theory of, 132; semi-independent, 18; token-token scheme and, 38–39, 46–48, 50
brain state, open question argument, 96–97, 99
Brainstorms (Dennett), 40
Brett, R. L., 56, 216n4
Brink, David, 96
Buddhist thought, 136
buffer-store, 87–88
Burge, Tyler, 104–5, 189
Butler, Bishop, 125–26
Byers, Sarah, 67–68

Cambridge Platonists, 53
Carruthers, Peter, 87–88
Cartesianism, 156, 182, 194; as "elderly," 104–5, 135; HOT theories and, 83, 85; misleading characterizations, 7, 15, 57, 76–77, 81; modern view of, 62; two-substance ontology, 52–53, 66, 102, 219n47
Cartesian theater, 17, 106
Cartwright, Nancy, 118
Caston, Victor, 9
causation, 170; backward, 196–97; executive, 120–21; information and reason, 81; mental, 14, 24–25, 77–78, 130, 170, 188–90, 198; mind-body interaction and, 78–80, 184–90; motive and, 183–84; nonphysical modes of, 22–23; personal identity and, 123–24, 130; regularity theory of, 191–96; seventeenth-century views, 54–55
Chalmers, David, 36–37, 91, 206–7, 211n1
characterological psychology, 175
"Chinese Room" criticism, 42–43
Christianity: academic philosophers and, 57–58; eschatology, 114–15
Cicero, 62
civic life, 208
Clarke, Desmond, 62
classification, 116–17, 120
clear and distinct perception, 64–66
clinical epistemology, 166–67
Cogito, 63–66; Augustine's anticipation of, 57–58, 217n16; as device against skepticism, 58, 60, 101–2; moral form of, 144
cognitive neuroscience, 7; Cartesian features of, 81, 83; emotion and, 159–60; folk psychology as foundational, 16, 208–9; limitations

of, 198–99, 202, 209; lived life and, 206–8
cognitivism, 155, 163–64, 168
common notions, 63, 129
common sense, mechanism and, 52–53
communists, 77
comparative psychology, 28
computer model of mental life, 39–40, 41, 42–43
Concept of Mind, The (Ryle), 20–21
Conceptual Role Semantics (CRS), 42
consciousness: Augustine on, 57–58; awareness of mental states, 86; as feature of reality, 19, 36; as function of brain, 27–28; higher-order thought required, 84; introspective, 45–46; lack of role for, 1–2; planning and, 201–2; present in animals, 59; quantum, 23–24; reduction to physical state, 27–28; as shared knowledge, 20–21
consciousness, problem of, 1–3, 15, 60; consciousness as special kind, 26–27; essentialism and, 18–19; Greeks and, 2–16; HOT theories and, 88; mental-physical relationship and, 80. *See also* experience
Consciousness Explained (Dennett), 17–18, 40–41
constancy phenomena, 41
content: nonmental, 89–90; states and, 89–90
context, 79
continuity, 96; of identity, 111–13; of memory, 125; psychological, 126–27
Contradiction, Law of, 146
core consciousness, 156
covering laws, 79
Crane, Tim, 74, 188, 190
credibility, 33, 63, 85, 171
Critique of Pure Reason (Kant), 143–44

Damasio, Antonio, 155–57
Davidson, Donald, 78–81
decision making, 147–48, 165–67
De Corpore (Hobbes), 110
deliberation, 7, 59, 148–50, 164, 170–71, 174, 201, 205
democratization, 77
Democritus, 5
Dennett, Daniel, 17–18, 37, 40–41, 203–4
Descartes, René, 15; doubt, method of, 85; educational background, 61–63, 67; on externals, 105–6; letters to Princess Elizabeth, 184–88; ontological blunder of, 27; physicalism of, 188–90; probabilities, rejection of, 62–63; theological principle, 64; *Works: Discourse on Method*, 55–56, 62–63, 85, 105–6; *The Passions of the Soul*, 76, 157–58; *Principles of Philosophy*, 64
Descartes' Error (Damasio), 156
descriptions, 118–19
determinism, 199–200
De Trinitate (Augustine), 57–58
Dewey, John, 76–77

dichromatism, 46, 106–7, 109
directed awareness, 202
Discourse on Method (Descartes), 55–56, 62–63, 85, 105–6
discovery, 54, 64, 103, 107, 193
Diseases of the Personality (Ribot), 132–33
dissociative conditions, 131–34
"Does 'Consciousness' Exist?" (James), 97–98
doing, awareness and, 86
doubt, method of, 85
Dretske, Fred, 94

ego, dissociative conditions and, 132–34
Ekman, Paul, 149–50, 151
eliminativism, 27, 97
emergentism, 28, 30, 40
emotion, 2; anger, 179–80; Aristotle's treatment of, 158–59; attributed to brain, 150–51; charioteer metaphor, 145–46; cultural differences and, 151–52; decision making and, 147–48, 165–67; as dispositional, 154–55; as evaluative, 154–55; evolutionary model, 149–50, 152, 156, 160–61; explanation and, 161–63; facial expressions and, 149–50; feelings, 155–56; interests and, 148–49; law and, 167–68; lived life and, 159–61, 165, 167–68; moods, 150; motives and, 178–79; neuroscience and, 159–60; personality and, 155; propositional attitude and, 164–65; psychology of, 159; reason and, 152–53, 163–64; as sensation, 154–55
emotivism, 155
empathy, 208
Empedocles, 4
emphasis, 142–43
empirical methods, 65, 68, 119, 136, 143
Enquiry (Hume), 194–95
epideictic rhetoric, 174
epiphenomenalism, 33–34
epistemic credulity, 118–19
Essay Concerning Human Understanding, An (Locke), 20, 113–16
essence: Aristotle's view, 116–17; nominal, 111, 113–14, 117–18; real, 111, 113–14
essentialism, 18–19, 30, 116–22; psychological, 119–21
ethological perspective, 173
Euclid, 129
evolution, 101–2
evolutionary model of emotion, 149–50, 152, 156, 160–61
evolutionary psychology, 28–29, 96, 160–61
executive causation, 120–21
experience, 91, 135–37; bat example, 44–45, 91; consciousness as everyday fact of, 9–10; as fundamental, 36–37; monistic ontology, 97–99; subjective, 208–9
explanation: clarification vs., 68–69; emotion and, 161–63
explanatory gap, 34–35, 37, 158, 204
externalism, 104–11

facial expressions, 149–50
Fechner, Gustav, 94
feelings, 155–56
filtering, 202
first-order thought, 84, 92, 94
first-person accounts, 21 to, 25, 46, 65, 96–97; awareness and, 86; HOT theories and, 93
first-person problem, 206–7
fitness for the rule of law, 117
Flourens, Pierre, 132
folk psychology, 65, 86, 168; as foundational science, 16, 208–9; planning and, 202–3
forensic rhetoric, 174, 177–78
Foundations of Human and Animal Emotions (Panksepp), 150–51, 227n7
French psychologists, 131–32
functionalism, 37–43, 113, 186; metabolism and, 205–6; radar equation, 39–40

Galileo, 53
Gall, Franz, 132
Garver, Eugene, 171–72
Gassendi, Pierre, 28, 214n15
ghost in the machinery, 75–76
Gilovich, Thomas, 166
Goldie, Peter, 155
Goldman, Alvin, 84
Graham, C. H., 107, 1091
Greeks, problem of consciousness and, 2–16

habit, 195–96
Hacker, Peter, i46
Hamlet (Shakespeare), 152, 153
harmony theory, 4–6
Harré, R., 138
Hawthorne, John, 130
Heidegger, Martin, 72–73
Heil, John, 186
Hempel, Carl, 22
Herbart, Johann Friederich, 94
higher-order thought (HOT) theories, 83; animals and, 85, 89, 92–93, 95; Cartesian elements in, 84–85; HOT-d and HOT-nd, 87–88; mental states and, 86–87; as metacognitive theory, 84; unclear assumptions and phrases, 86; versions of, 87–88
Hobbes, Thomas, 20, 28, 111–13, 216–17n9; critique of, 53–54
homunculi proposal, 40–41
How We Know What Isn't So (Gilovich), 166
Hsia, Yun, 107, 109
Hubler, J. Noel, 10
Hume, David, 122–24, 130, 140, 171, 183–84; regularity theory of causation, 191–96
Huxley, Thomas Henry, 33–34
hypotheses, 34–35

identity, 37–39; token-token, 38–39, 50; type-token, 38, 46–47; type-type, 38, 46–47
identity, personal, 111–16; continuity of, 111–13; inner self, 127–28; laws of association and, 122–24; memory and, 114, 122, 124–25; ownership and, 141; Prince and

identity, personal (*continued*)
the Cobbler example, 114, 122; self-identity and, 140–41; starting point and, 129, 131. *See also* self; self-consciousness
Iliad (Homer), 145
imagination, 8, 187, 202
individualism, 104–5
inference, habit of, 195–96
information, reason and, 81
intentionality, 87–88, 203
intentional objects, 73–75, 187
interests, 148–49
introspective consciousness, 45–46
intuition, 129, 130, 196
invention of mind, 71–72

Jackson, Frank, 106–7
James, William, 2, 89–90, 97–99, 156; determinism, view of, 199–200; experience as fundamental, 36–37; substance theory of self, 135–36, 139–42; *Works:* "Does 'Consciousness' Exist?" 97–98; *Principles of Psychology*, 23–24, 90
Jesuit schools, 61–63
judgment, 7, 65, 170–71
justice, emotion and, 167–68

Kant, Immanuel, 143–44
Kenny, Anthony, 59, 155
Kim, Jaegwon, 29–30
knowledge: consciousness as shared, 20–21; motive and, 185–86; no extent of knowledge provision, 30–31; reflexively transparent, 131
Korsgaard, Christine, 128
Kripke, Saul, 118

La Fleche, 61–63, 67
La Mettrie, 28
language: Augustine's view, 58–59; limits of, 68–70; natural, 71; pronomials, 142–43
Laplace, Pierre Simon, 54–55
layered conception of reality, 32
Lehrbuch der Psychologie (Herbart), 94
Leibniz, Gottfried, 74
Levine, Joseph, 34–35
Lewis, David, 189
"Life of Theseus" (Plutarch), 111
lightning example, 37–38
linguistic analysis, 15, 68–70; intentional objects and, 73–75
linguistic turn, 68
lived life: cognitive neuroscience and, 206–8; emotion and, 159–61, 165, 167–68
Livingston, Paul, 70
Lochwood, Michael, 23
Locke, John, 20, 53, 111, 113–16, 122, 140
logical positivism, 79
lower-order thought (LOT), 89, 95

Mackie, J. L., 196
magnitude of response, 49
magnitude of sensation, 49
Mary problem, 106–7
materialism, 57, 77, 208; antimaterialism as alternative to, 34–35
mathematics, 14, 63, 129

Matson, Wallace, 3, 8, 13, 14
meaning, 42, 69, 103; qualia and, 106–7, 109; Twin Earth example, 104–10
"Meaning of 'Meaning,' The" (Putnam), 104–10
mechanism, 52–53, 56
memory, 211n3; personal identity and, 114, 122, 124–25; quasi-memories, 125–26
mental causation, 14, 24–25, 77–78, 170, 188–90, 198
mental disorders, 91, 131–34, 135
mental states: brain function dependent on, 23–24; HOT theories and, 86–87; nonconscious, 89; subjectivity, 44–46
mereological fallacy, 146, 165
metabolism, 67–68
metacognitive state, 84, 86
metaphysics, 14–15, 56
methodology, philosophical, 55–56
Metzinger, Thomas, 101
mind: as conceptual space, 8–9; invention of, 71–72; naturalization of, 94–95, 99
mind-body relation: Aristotle's view, 2–12; causation and, 78–80, 184–90; historical position, 2–14, 51–52; motive and, 184–88; Plato's view, 12–14
mind dust theories, 89–90
Minsky, Marvin, 138
Mintz, Samuel, 53–54
models, 101–2
modularity theories, 6–7, 40, 132
monism, anomalous, 78–81
monistic ontology, 97–99
Moore, G. E., 96, 99
Moral Order and Progress: An Analysis of Ethical Conceptions (Alexander), 29
moral philosophy, 169–71
moral properties, as nonnatural, 96, 99
More, Henry, 53
Morgan, C. Lloyd, 28–29, 214n17
motive, 2, 170–71; causation and, 183–84; character and, 177–82; character of speaker and, 172–77; emotion and, 178–79; forensic rhetoric and, 177–78; inseparable from persons, 198–99; knowledge and, 185–86; mind-body interaction and, 184–88

Nagel, Thomas, 44–45, 91
narrator, 122
National Socialism, 73
natural, incomplete ontologies of, 99–100
naturalization of mind, 94–95, 99
natural kinds, 117–18
natural slave, 117, 182
neopragmatism, 77
neural processes, 29, 189
neurological disease: blindsight, 89; phantom limb, 89, 183
Newtonian theory, 115–16
no extent of knowledge provision, 30–31
nominal essence, 111, 113–14, 117–18
nonconceptual content, 87–88
nonconscious mental states, 89

nonmental content, 89–90
Norman, Donald, 158

observation, 118–19
On Double Consciousness (Binet), 131–32
"On the Hypothesis That Animals Are Automata" (Huxley), 33–34
On the Soul (Aristotle), 4, 9–10
ontology: monistic, 97–99; two-substance, 52–53, 66, 102, 219n47
open question argument, 96–97, 99
ownership, 42–45, 141

pain example, 69, 137, 186–87
Panksepp, Jaak, 150–51, 227n7
panpsychism, 24
Papineau, David, 22
"Paralogisms of Pure Reason" (Kant), 143
Parfit, Derek, 124–28
PARFITON, 125
Passions of the Soul, The (Descartes), 76, 157–58
perception, 3, 154, 216n41; Aristotle's view, 7–8, 10–11; Augustine's philosophy, 57–58; clear and distinct, 64–66; person and, 122–24
periodic table of elements, 99–100
personality, emotion and, 155
persons, 122–24
persuasion, modes of, 171–73
Phaedrus (Plato), 145–46
phantom limb example, 89, 183
phenomenology, ownership and, 42–45
Philosophical Essay on Probabilities (Laplace), 54–55
Philosophy and the Mirror of Nature (Rorty), 71
physicalism, 14, 30, 67, 188–90; as alternative to antimaterialism, 34–35; commitment of cognitive neuroscience to, 83; defenses of, 22–25, 81; introspective consciousness, 45–46
physics as complete, 13–14, 22, 60, 212n24, 213n6; higher-order thought and, 85; seventeenth-century views, 54–55
physiology as complete, 22–23
pineal gland, 157
planning, 201–2; computers and, 204–5; folk psychology and, 202–3
Plato, 2–3, 145–46; mind-body problem and, 12–14
Plutarch, 111
Politics (Aristotle), 182
pragmatism, 121–22
Prince and the Cobbler example, 114, 122
Principle of Causal Interaction, 78, 79–80
Principle of Nomological Character of Causality, 78
Principle of the Anomalism of the Mental, 78
Principles of Philosophy (Descartes), 64
Principles of Psychology (James), 23–24, 90
privacy claim, 69–70

private-language question, 137
probabilities, 62, 63, 147
processes, 21
propositional attitude, 164–65
psyche, 3
psychological continuity, 126–27
psychological essentialism, 119–21
psychophysical relations, 79
psychophysics, 93–94
Putnam, Hilary, 104–10

qualia, 31, 49; meaning and, 106–7, 109
quantum mechanics, 23–24
quarks, 100
quasi-memory, 125–26

radar equation, 39–40
rational choice, 6
rational decision making, 148
rationality, 59–60, 95; emotion and, 146–47
Ratio Studiorum, 61–63
real essence, 111, 113–14
reality, 19, 32, 36
reason, 131; emotion and, 163–64
reasonableness, 147–48
Reasons and Persons (Parfit), 124–28
recipient bodies thought experiment, 124
reductionist strategies, 44–46, 146
reflexively transparent knowledge, 131
regularity theory of causation, 191–96
Reid, Thomas, 72, 185, 196–98
representation, 8, 42
res cogitans, 27, 70, 85, 121; externalism and, 105–5
res extensa, 27, 65, 121
retinal cone cell example, 90–91
rhetoric, 170–71; as applied psychology, 170–71; branches of, 174; character of speaker and, 172–77; reasoning and, 171–72
Rhetoric (Aristotle), 159, 171–82
Ribot, Theodule, 131–32
rich inductive potential, 120–21
Robinson, T. M., 13
"Romantic rebellion," 72–73
Rorty, Richard, 71–73, 139
Rose, Hilary, 161
Rose, Stephen, 161
Rosenthal, David, 220n1
Ryle, Gilbert, 20–21, 75, 111

Scholastic philosophy, 52, 56, 59, 61–63
scientific realism, critiques of, 118
Scottish Common Sense Philosophy, 72
Searle, John, 42–43
self: belief in, 138; inner, 127–28; as model, 101–2; personal identity and, 111–16; social construction of, 110–11, 134–43; socioempirical, 127–28; substance theory of, 135–36, 139–44; will and, 143–44. *See also* identity, personal
self-consciousness, 25, 101–2; dissociative conditions and, 131–34; essentialism and, 116–22; externalism and anti-individualism,

self-consciousness (*continued*) 104–11; meaning and, 103; spatiotemporal conditions and, 102–3, 130. *See also* identity, personal
selfhood, 141
self-identity, 140–41
self-movement, 67–68
sensation, 9, 12; emotion and, 154–55; functionalist approach, 37–38; terms for, 69–70
senses, 7–8, 58
Series in Affective Science, 150
Shaftesbury, Earl of, 55
Ship of Theseus problem, 111, 112
Siewert, Charles, 65, 87
Signal Detection Theory (SDT), 221n8
simulacrum theory, 118
skepticism, 21; *Cogito* as device against, 58, 60, 101–2
sleep, consciousness during, 26
social constructionism, 110–11, 134–43
Socrates, 181
soft determinism, 199–200
"Solidarity or Objectivity?" (Rorty), 139
Solomon, Robert, 155
soul: modularity theories, 6–7; as principle of action, 4–5, 7; self and, 143–44; as source of all movement, 13–14, 67–68
"Soul and Body" (Dewey), 76–77
Space, Time, and Deity (Alexander), 29
spatial states, 73–75
spatiotemporal conditions: self-consciousness and, 102–3, 130
speaker, character of, 172–77
stability, 46–47
Staden, Heinrich von, 57
starting points, 14, 63, 129
states, 21, 33; consciousness of vs. being conscious of, 87; contents and, 89–90; functional, 113; higher-order thought theories and, 86; metacognitive, 84, 86; nonconscious, 89; spatial, 73–75. *See* also mental states
Stillingfleet, Edward (Bishop of Worcester), 115
stipulations, 22, 63, 185–86
Stoic philosophy, 115, 153
subjectivity, 44–46; of decision making, 147–48
substance theory of self, 135–36, 139–44
supervenience theory (ST), 29–33

teletransportation thought experiment, 125
third-person account, 207, 209
thought: first-order, 84, 92, 94; imprecision of concept of, 92; lower-order (LOT), 89, 95; as neural discharge, 38
tiger example, 118, 156
tip-of-the-tongue phenomena, 84, 86, 87
token-token identity, 38–39, 50
Treatise of Human Nature, A (Hume), 122–24

truth, 10
Twin Earth example, 104–10
Two Treatises of Civil Government (Locke), 115–16
type-token identity, 38, 46–47
type-type identity, 38, 46–47

unity of consciousness, 24–26
universal categories, 115–16

van Fraassen, Bastian, 118
verificationism, 68
virtue ethics, 169
visual perception example, 48–49
vitalism, 37
volition, 98, 193, 197, 202

water, shifting meaning of, 109–10
water properties example, 30–31
"Why Isn't the Mind-Body Problem Ancient?" (Matson), 3, 8, 13, 14
Why Utility Pleases (Hume), 171
Wiggins, David, 124
will, 143–44
Windelband, Wilhelm, 219n42
witness, 98–99
Wittgenstein, Ludwig, 69–70, 102, 105, 137–38